The Experimentation Field Book

The Experimentation Field Book

A Step-by-Step Project Guide

BY

Jeanne Liedtka

Elizabeth Chen

Natalie Foley

David Kester

Columbia Business School
Publishing

Columbia University Press
Publishers Since 1893
New York   Chichester, West Sussex
Copyright © 2024 Jeanne Liedtka, Elizabeth Chen,
Natalie Foley, and David Kester

Cataloging-in-Publication Data
available from the Library of Congress
LCCN 2023030499
ISBN 978-0-231-21417-9 (paper)
ISBN 978-0-231-56022-1 (electronic)

Printed in the United States of America

Book design by Leigh Ayers, 10/HALF Studios

Contents

Introduction to the Experimentation Process

DOES THE WORD *experimentation* conjure images of test tubes, safety goggles, and possibly a dash of fire or smoke? Our early exposure to experimentation is often in science class as part of the scientific discovery process. Accompanying those beakers and lab coats is the scientific method, a guide to running experiments.

But many people run experiments every day outside of science labs. Long before you were learning about experiments in school, your toddler brain was experimenting as a primary method of learning: *If I stack blocks like this, they don't fall over. If I put my fingers near the door, I can get hurt when it shuts.*

Many organizations use experimentation to explore ideas for new products, services, programs, and strategies. In this field book, we'll share stories from different **organizations**. Nike wanted to experiment

Organizations in Brief

 Nike is one of the largest and most successful athletic apparel companies in the world.

 Whiteriver Hospital, located on the Fort Apache Indian Reservation, has both inpatient beds and an emergency room where most of the population's medical needs have to be met.

 South Western Railway (SWR), a joint venture between two of the world's leading rail companies, operates some of the busiest train routes in the United Kingdom, with 235 million passenger journeys a year.

 The Project Management Institute (PMI) is the premier global project management association, with nearly 500,000 members.

with a shoe subscription service for kids. Whiteriver Hospital wanted to reduce patient wait time in their emergency room. The Project Management Institute (PMI), a professional learning organization, was interested in testing different ways to serve its members. South Western Railway (SWR), one of the UK's biggest commuter railways, hoped to improve passenger experiences. People at each of these organizations had good ideas on paper but wondered if they would succeed in reality. They worried about being wrong in their hunches. So, they used experimentation to test their hunches *before* they built and launched their new ideas and *before* spending a lot of time and money on them. Throughout this field book, we will return to these four stories for inspiration and lessons learned as we build your experimentation skills.

Entrepreneurs, teachers, caregivers, marketers, graphic designers, social workers, even people doing their everyday routines all experiment. Have you ever caught yourself saying, "I wonder if taking this route home is faster? Let me try it and time it." You're experimenting! Or think of a team leader considering, "If we send the agenda in advance, it might make our meetings more effective. Let's try it for a few weeks and see if that's true." Experimenting! This field book focuses on helping you to systematically design and execute your own high-quality experiments on any idea you have.

Experimentation is much talked about in the abstract, but poorly understood in practice. Yet it is the crucial connector between *imagining* an idea and *making it happen* in reality.

Why Experiment?

Here are a few reasons putting in the time to experiment is better than just building a new idea:

- Experimentation protects you from *overspending* on a solution that won't work for the people you designed it for. By placing small bets and learning at a fast pace, you can learn whether your concept really fulfills the unmet needs of your users—whether patients, passengers, customers, or members—and if the idea is really desirable to them.

- It encourages you to test a *portfolio* containing multiple ideas rather than converging prematurely on a single idea. Then it lets your users highlight the one that works best for them.

- Through experiments, you learn how to *scale* your idea effectively and successfully. Testing aspects of the idea at smaller scales reduces the risk of early service delivery friction and is less expensive.

- You'll get to know your *early adopters*, and their feedback will inform future features and help accelerate the development of your new product or service.

- Experimentation is fun and *engaging*! It gets you out of the conference room and into the real world and allows you to invite those who will be part of the implementation of a solution into the testing process, building buy-in and adding energy and momentum to a project.

Experiments exist to collect the data to build an evidence base, to support whether and how to launch new solutions, but they are often underutilized or badly designed. At one extreme, they end up looking like "soft launches" with little appetite for pivoting. In other fields, only elaborate, formal experiments like randomized controlled trials, which often last years rather than days or weeks, count. In this book, we will show you how to take advantage of the often-unexplored territory in between, where well-designed, learning-oriented field experiments can add tremendous learning and risk reduction but take only hours instead of weeks and cost next to nothing.

For most of us, building an evidence base is the best way to manage the inevitable risk of trying new things in today's dynamic environments. In fact, skills in the design and execution of experiments are fundamental and critical competencies for success in an increasingly uncertain world where there are known unknowns (things we *know* we don't know) and unknown unknowns (things we *don't* know we don't know). Experimentation helps us to better address the first category and discover the second one. This valuable tool is embedded into many problem-solving approaches (like Lean Startup, Agile Software Development, Design Thinking, Kaizen, and Process Improvement). Experiments also go by many names; you might have heard about a new Google feature that is "in beta" or about organizations running pilots or testing a minimum viable product (MVP). Yes, the terms used for experimentation can be confusing and full of jargon—but the terms we use are less important than being precise about what the experiment seeks to test and how it will accomplish that. Experimentation is a powerful and effective learning tool for individuals or teams who are improving a product or service that exists or creating something new to them, their organization, or the world—all require deliberate learning through action. Though the need for experimentation may seem obvious, many organizations skip it. Why? Because the urge to "just build it" (and hope they will come) is strong.

A hypothesis is the essential starting point for experimentation. Yes, back to science class! A hypothesis is a best guess about what you expect to be true or assume is true. Nike employees could have just assumed parents would want a shoe subscription service and built it, but instead they experimented to test the assumption that it met a real customer need.

Hypotheses are conjectures that can be true or false. We find out if they're true by collecting data to confirm or disconfirm them. For example, you can test if that new route you suspect to be faster truly is by timing it and then comparing that to your old route time. Hypothesis-driven thinking is the anchor of experimentation—and it is fundamentally different from just using existing data to make your decision. Hypothesis-driven thinking flips the usual process that starts with the data you've already got. Instead, you start with the hypothesis and then identify the data you need to test it rigorously (which is often *not* the data you've already got). For your data to be most useful, your data collection must be done with intention. To do that, you need to be hypothesis driven.

Experimentation is not a "one and done" activity. It is a process that evolves over time. You will run multiple tests, which will increase in complexity as your hypothesis is further developed and refined. This de-risks your solution development. Rather than just building the real thing right away, experimenting helps you manage risk (you spend time and money incrementally as risk moves from higher to lower), gather evidence to test your assumptions from a range of actual users (not only from conversations among colleagues or early adopters), and improve your odds of success. The iteration that occurs with experimentation creates a stronger solution by making sure that people really want what you've designed and demonstrating that you can deliver it in a viable way. Experimentation can also be used to understand whether any given problem is even worth solving.

We hope that we've convinced you of the *why* of experimentation and the need to be hypothesis driven. In the rest of this field book, we will focus on the *how* by providing a hands-on, structured process to guide you through the design and execution of your very own experiments. We will outline how to design and run tests to address a variety of challenges—improving an existing solution, fixing something that's broken, making a new product, or launching a new service for your existing customers. To do this, we provide a step-by-step process, using the templates we have created. Together we will explore **five steps**.

Let's look briefly at each step. Steps 1 and 2 focus on the *what* of experimenting. Step 1, framing a testable idea, begins the process. Because we will treat our ideas as hypotheses, we need to consider what makes any given

The Five Steps

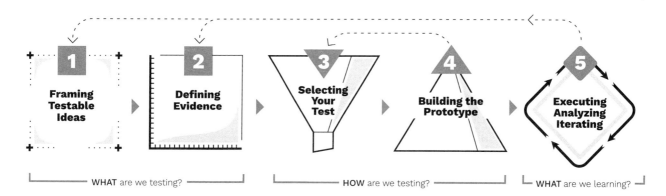

idea truly testable. We will explore this as we look at the specifics of your idea and who it serves—information that is needed to design a rigorous test. Next, in Step 2, we define what constitutes evidence—what kind of data will tell us whether our hypothesis (and the assumptions behind it) is true or false? Where will we locate such data? Having specified what we are looking for, we then move on to the question of *how* to conduct our experiment. In Step 3 we sort through a variety of options to zero in on the best type of test to collect the desired data. What type of test best suits the particulars of our idea and the evidence we need to gather? Once we have selected the test type, in Step 4 we develop the stimulus we will use to provoke feedback by building the prototype, the simplest one that will do the job.

Prototypes and Pyramids

Prototyping has been around a long time! The Great Pyramid at Giza, constructed in 2528 BC, is the oldest of the Seven Wonders of the World and the only one that remains largely intact. Scholars believe that its construction, estimated to take decades—or even centuries—almost certainly relied on an iterative process using low-fidelity prototypes, including drawings and schematics. "The wonders of ancient Greece and Egypt required engineering far in advance of their time, and marvels of engineering such as the Pyramids and the Parthenon didn't sprout up out of the ground whole," explains engineering expert David Walsh.[*]

[*] "The Top Four Ancient Design Prototypes" by David Walsh, 8/19/15, https://www.asme.org/topics-resources/content/top-4-ancient-design-prototypes

Finally, with the key ingredients in hand—our testable idea (our hypothesis), our test type (our intervention), and our prototype (our stimulus)—we move on to the question of what we have *learned* in Step 5, as we execute our test plan, analyze the results it produces, and iterate our way to an improved solution.

But these steps are rarely as linear as our model suggests. Frequently we loop back to earlier steps. The selection of a test type, Step 3, and prototype format, Step 4, are usually highly interactive, and it is not uncommon to move back and forth in an iterative way to complete them. Sometimes our ideas fail in Step 5 and we loop back to Step 1 to start all over again with a new idea. Or maybe they succeed, so we loop back to Step 2 to design a more sophisticated test with a higher-fidelity prototype and different evidence—or to test a different set of assumptions. And, of course, things happen *before* our Step 1—the process of generating the idea, for instance—while the essential activities around scaling an idea in the real world occur *after* our Step 5, when the period of experimentation ends and implementation begins.

As we move through these five steps and apply them to your idea, we will illustrate them in action with examples from our own experiences and research working with innovators at Nike, Whiteriver Hospital, the Project Management Institute, and South Western Railway. Because the ideas you will want to test come in different forms—products, services, processes, software, or a combination of these—our stories encompass this variety. Let's preview these four stories, which we will return to time and time again.

Our Four Stories in Detail

Nike *Easykicks*

Context

Nike is one of the largest and most successful athletic apparel companies in the world. Like many consumer packaged-goods firms, they wanted to explore expanding into services that would allow them to diversify in ways that would develop deeper customer relationships.

Challenge

Nike had identified some problems worth addressing: kids grew fast and wrecked shoes, so parents constantly needed to buy new ones to accommodate their kids' growing feet. Shopping at a store with kids wasn't typically enjoyable. In addition, though some old shoes were donated, many old pairs lay abandoned in musty closets, and Nike, with a commitment to sustainability, had invested in technology to recycle old shoe materials. So, they wanted to explore the idea of a shoe subscription service, *Easykicks*, for young athletes, a population they wanted to get to know and serve better. The new subscription service offered an opportunity to address both the hassle of shoe shopping and the need for recycling.

Process

Nike knew they needed help in testing the business model of services, like *Easykicks*, that had caused challenges for them in the past, so they partnered with Peer Insight, a firm known for acumen in market experimentation. Peer Insight led a process using the steps outlined in this field book to design and run tests on the new subscription service over an 18-month period.

Whiteriver Hospital

Context

Whiteriver Hospital is located on the Fort Apache Indian Reservation, which covers more than a million acres and serves a population upward of 15,000, mostly Apache. It falls under the jurisdiction of the U.S. Department of Health and Human Services (HHS) and has both inpatient beds and an emergency room where most of the population's medical needs have to be met, including prescription refills.

Challenge

The hospital faced a serious situation: close to 25 percent of emergency department (ED) visitors were leaving without being seen, a problem attributed to long wait times. Nonemergency patients consistently got delayed as staff addressed true emergencies, with arrivals sometimes waiting as long as six hours before being seen. When potential patients left the emergency room (which they did at a rate twenty times the U.S. national average), their medical problems worsened. Often these eventually became true emergencies, and patients needed to be helicoptered off the reservation for more extensive and expensive care. The Whiteriver Performance Improvement Team wanted to explore the idea of adding an *Electronic Kiosk*, similar to one at Johns Hopkins Hospital in Baltimore, where a patient electronically signed in upon arrival, and the electronic system informed other parts of the hospital of that patient's potential needs, saving administrative time and speeding up the intake process for patients so that more patients could be seen more quickly in the ED.

Process

HHS operated the Ignite Accelerator, a program of their IDEA Lab aimed at bringing new innovation approaches to employees across the United States. Ignite offered education, coaching, and a small funding stipend, to boost projects that offered the hope of addressing agency problems. Whiteriver's Performance Improvement Team was invited to test the electronic kiosk idea as part of the Ignite program, using a process similar to that outlined here.

Our Four Stories in Detail (*continued*)

South Western Railway

Context

SWR, a joint venture between two of the world's leading rail companies, operates some of the busiest train routes in the United Kingdom, with 235 million passenger journeys a year. SWR faced challenges with industrial relations, major network repairs, staff morale, and the passenger experience.

Challenge

SWR leadership saw an opportunity to improve the passenger experience. Unlike Nike and Whiteriver, SWR did not start with a solution in mind; they wanted to first explore the problems passengers faced in more detail. They commissioned the consultancy David Kester and Associates (DK&A), a leading design firm, to help them find both quick wins and a long-term strategy to improve the SWR customer experience. They captured the challenge: to learn fast from customers, rapidly deliver confidence-building basics at SWR stations, and together shape the future SWR experience.

Process

Using ethnographic research tools, DK&A partnered with SWR staff at three central rail stations to better understand the problems and needs of their passengers. They synthesized this research, identifying insights and creating journey maps and personas. Based on these, they facilitated a collective brainstorming process that identified multiple ideas. Three concepts emerged as particularly attractive: the *Concierge* (aimed at offering a warm welcome and providing the right infrastructure to better support customers in the ticketing process), *Cleaner and Cleaner* (a hygiene awareness and nudge campaign to provide reassurance to the public), and the *Wayfinding Audit* (addressing poor signage and visual clutter with smart customer-led tools). DK&A then embarked on an experimentation phase, using a process similar to the steps outlined here.

Project Management Institute

Context

PMI is the premier global project management association, with nearly 500,000 members. Like many membership associations, they struggle with how to evolve and continue to add value for their members beyond their core product, the Project Management Professional (PMP) certification.

Challenge

PMI had already done extensive research to better understand the problems and needs of their members, and had identified four high-level ideas they thought had significant potential to create value for them:

1. *Snippets*: A microlearning platform that curated quick, bite-sized trainings and resources to fit members' day-to-day learning needs.
2. *Career Navigator*: A self-assessment tool that showed potential career paths based on a member's experience, skills, and interests.
3. *Hive*: A peer-to-peer connection platform that allowed members to get answers to their toughest project management questions by connecting directly to experienced peers.
4. *Spot*: An experiential learning opportunity that matched members with real-world, low-risk opportunities to help them hone needed skills.

Process

Each idea was intriguing but untested, and with an important board meeting approaching, PMI leadership needed to make choices about which ideas to invest in. They partnered with Peer Insight to run a series of experiments over three months on the four offerings to narrow down which they should move forward. From there, Peer Insight ran a series of deeper market experiments over six months on the best performing concepts, again using our five-step process.

Each of these organizations spent time to first understand the problems their customers faced, to ensure that they invested in solutions that solved problems that mattered. We will pick up their stories as each enters the experimentation process with particular ideas that they want to test. Though experimentation can be used to identify problems as well as to test solutions, in this field book we will focus on how to test a given solution—like *Easykicks*, the *Electronic Kiosk*, the *Concierge*, or *Career Navigator*—to see whether it does, in fact, meet an identified need.

As we get started, a note to you, our reader: we wrote this book under the assumption that you have an idea of your own that you want to test. Great! This is the most effective way to work with this book—to learn while doing—whether it's with an actual idea or one you want to use just to learn with. We will use your ideas to practice the learning-by-doing process in Steps 1 through 5. In each step, we will give you an assignment to apply what we've talked about to your own idea—so you'll be both learning and doing as we go. We will ask you to note the milestones in your journey through the five steps using our **Progress Tracker** [TEMPLATE 1, PAGE 82].

By the time we conclude, you will have followed your idea through one full cycle of testing.

Let's get started exploring the world of experimentation!

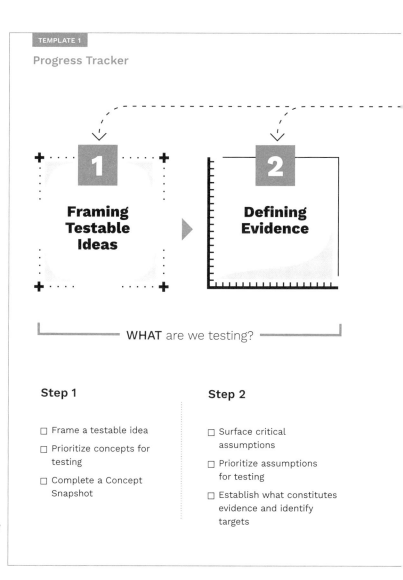

TEMPLATE 1

Progress Tracker

1 Framing Testable Ideas

2 Defining Evidence

WHAT are we testing?

Step 1

☐ Frame a testable idea
☐ Prioritize concepts for testing
☐ Complete a Concept Snapshot

Step 2

☐ Surface critical assumptions
☐ Prioritize assumptions for testing
☐ Establish what constitutes evidence and identify targets

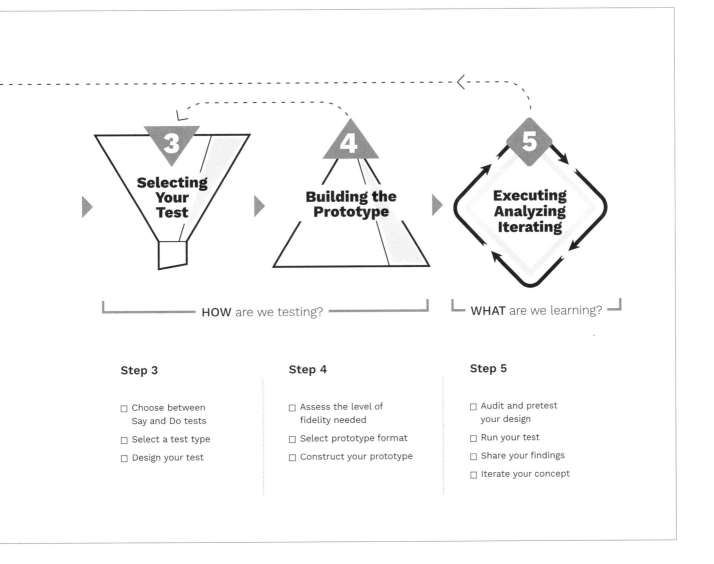

Selecting Your Test (3)

Building the Prototype (4)

Executing Analyzing Iterating (5)

HOW are we testing? ——— **WHAT** are we learning?

Step 3

☐ Choose between Say and Do tests
☐ Select a test type
☐ Design your test

Step 4

☐ Assess the level of fidelity needed
☐ Select prototype format
☐ Construct your prototype

Step 5

☐ Audit and pretest your design
☐ Run your test
☐ Share your findings
☐ Iterate your concept

On to Step 1!

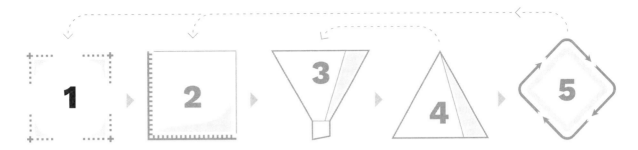

STEP 1

Framing Testable Ideas

IN STEP 1 YOU WILL LEARN TO:

- Frame a testable idea

- Prioritize concepts for testing

- Complete a Concept Snapshot

Framing a Testable Idea

An idea that cannot be tested is useless; if you can't bring data to bear on it, it needs more development. Of the many supposed "hypotheses" generated by the learners we work with, 90 percent are *not* testable: they are vague and abstract. An idea for a new product, for instance, is by itself not testable. Testability requires identifying a target user, their needs, and specifics about how the solution will be executed. To be testable, your idea must also contain enough detail to make it clear to others.

Let's say that your friend has an idea for a tiny drone that hovers above a reader's head and shines a light on their book while they are reading at night in bed. That's not a testable idea . . . yet. It needs more definition: What kind of reader is the drone designed for? What problem is it trying to solve? How will it work? What makes it better than their current night light, lamp, or book light?

Let's look closer at these qualities that characterize a testable idea:

A testable idea is as *specific* as possible. Without sufficient detail, test results will be as vague as the idea you started with. You may be tempted to identify as broad a target market as possible for your new idea. Resist that impulse and focus instead on designing as simple and specific a test as possible for a clearly identified group of customers or users—in your friend's case, people who enjoy reading in bed at night. At Whiteriver Hospital, for instance, the team dug deeper beyond the broad category of "patients visiting the ER" to identify who the most frequent users were. They were tribal elders, many of whom did not read or speak English.

An idea worth testing also brings something *new* to the table. Technically, you could create and test an idea that looks just like existing offerings in the market. But why would you want to? Think about it: What kind of hypotheses did Sherlock Holmes like best? Never the obvious ones; you don't build your career as a famous detective—or an innovation leader—by proposing the obvious. It is fine for the idea to look obvious in retrospect (the best ones often do), but at the outset, it should feel *fresh*. In business parlance, we call these fresh ideas *differentiated*—distinct from what is already available to customers. Though the electronic kiosk that Whiteriver wanted to try wasn't new to the world, it would be new to their patients.

Finally, a good idea is *worth the trouble* to test. That is, the payoff should at least potentially be worth the effort in terms of producing a return for the organization as well as the user. Running experiments costs time and money. You want to invest your testing energy in places with a likelihood of a good payoff. Are you meeting the needs of a sufficiently large group of users? Is your idea deliverable at a reasonable cost? Will users be able to access and pay for the service as needed? Thinking about these longer-term issues doesn't mean that you expect all your ideas to succeed, but it does mean that those that do will be worth something. The Performance Improvement Team at Whiteriver believed that visitors departing without being seen were not only seriously risking their own health but also significantly impacting the hospital's financial health.

Often, the offerings you want to test are composed of several ideas. In our terminology, a concept refers to a combination of ideas. For example, *Easykicks*, Nike's subscription shoe service, is a concept that combines a few distinct ideas or components, such as creating a subscription service and recycling old shoes. To keep it simple, for the remainder of this book, we will refer to the hypothetical offerings we want to test, whether they are individual ideas or combinations of ideas, as *concepts*. We will capture their key elements in a **Concept Snapshot** (more on this coming up).

Prioritizing Concepts for Testing

In our introduction, we alluded to the power of moving multiple ideas into experimentation. By not "putting all of our eggs in one basket" and instead offering options to our customers (and then letting them choose), we create the most value for them at the least risk to ourselves. Accordingly, it's best to move a *portfolio* of concepts into testing. It is unlikely that you have the time and resources to test *all* of your concepts, so you need to have clear priorities about which to start with, to use your experimental energy well. Not all concepts are of equal value. In most cases, two dimensions are especially useful to think about in the prioritization process: (1) how much *value* or impact you believe the concept will create for the intended customer and

(2) how much *effort* it is likely to take to execute. The **Value/Effort Matrix** [TEMPLATE 2] uses these dimensions as axes to help you lay out and choose the particular ideas to include in your testing portfolio.

When using the matrix, you can also look for synergies among the different ideas, as quick wins can often be developed or combined to form more strategic concepts. The *Concierge* concept, for instance, was created by connecting three discrete ideas.

In selecting which concepts to prioritize for testing, the upper left quadrant (high value/easy to execute) is a great place to look for early wins. High-value, easy-to-execute concepts tend to be few but are logically a high priority for immediate attention. Conversely, low-value, hard-to-execute concepts (bottom right quadrant) are rarely of interest. However, most concepts will fall along a diagonal line with similar value-to-effort profiles. Think about your strategic goals as you select concepts to move into testing. Your prioritization process and criteria will differ depending on your projects' constraints, including time, resources, and risk tolerance.

This process, in theory, is similar to how you might choose stocks for an investment portfolio aimed at your retirement. If you're twenty-five years old, for example, you might orient your portfolio toward growth stocks with a high potential for capital appreciation. Conversely, at sixty-five, you would more likely play it safe, preferring less volatile ones with more stable returns. At forty-five, you might want a balance

TEMPLATE 2

Value/Effort Matrix

High Value Created
Easy to Execute

High Value Created
Hard to Execute

Low Value Created
Easy to Execute

Low Value Created
Hard to Execute

VALUE

EFFORT

between those two extremes. In the same way an investor assembles a stock portfolio with specific objectives in mind, you will assemble your portfolio of concepts for testing. Think about your larger strategic goals and priorities. Are you pressed for time with limited resources and looking for quick wins? Or, conversely, do you have an appetite for taking a higher risk profile to go for maximum visibility and impact? Or does a combination of payoffs to yield both short-term and longer-term results make the most sense? The answers to these questions depend on your individual strategy and circumstances.

Completing a Concept Snapshot

Once you have prioritized and selected your portfolio of concepts for testing, you want to ensure that your concept has the important characteristics— is specific, fresh, and worth testing—that we talked about earlier. The **Concept Snapshot** [TEMPLATE 3] helps you lay out a concise statement of your idea, specifying its key elements: the intended users, why it meets their needs, and why it is differentiated.

TEMPLATE 3

Easykicks Concept Snapshot

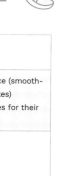

	User 1	User 2
For (target user)	**Young Athletes** (eight–twelve years old)	**Their Parents**
Who want (unmet needs)	• More control, autonomy, agency in their shoe shopping • The best shoes (providing the right fit, comfort, and performance) for various activities	• Convenient, more pleasant shopping experience (smoothing of shopping pain curve; smoothing the spikes) • Confidence that their kids have the right shoes for their activities
We will offer (offering)	A better, more positive shoe buying experience with: • An accurate fit assessment (digital or in-person) • A sustainable, cyclical buy-return process • A more attractive purchasing mechanism (subscription based)	
That provides (benefits)	• A creative outlet for self-expression • Confidence in their personal style • Special/unique feeling as they perform	• Shoes their young athletes need, when they need them • Less time spent shopping • A reduction in the pain of shopping (i.e., fewer conflicts, less stress)
Uniquely (differentiation)	• A personal (1:1), ongoing relationship between the firm and young athletes (build stronger relationships than created with products only; make young athletes aware of the brand earlier) • A service experience wrapped around shoes • Cyclical shoe experience • A *novel* business concept	

The Peer Insight team created a Concept Snapshot for *Easykicks* that highlighted two different users of the design (kids aged 8 to 12 and their parents), how it served their needs, what they would offer to accomplish this, and why the concept was differentiated.

To push concept development even deeper, you can create a simple storyboard. A storyboard is a visual tool used to show an experience through a series of pictures. Like a comic strip, it tells the story of your concept so that others can understand what you're imagining. It is one of the most basic forms of prototyping (we will talk more about this in Step 4). Despite its simplicity, the act of storyboarding is powerful for many reasons: It helps to flesh out a high-level idea. It also translates a concept from a static offering into a user experience. Creating this kind of "pre-experience" of a new idea is essential to getting accurate feedback from the people you are designing with and for. Finally, storyboarding helps a team align around a shared view of what they are creating together—you know that you are all talking about the same thing as you move into the testing process.

The Peer Insight team elaborated on their Concept Snapshot by storyboarding what the customer's experience using *Easykicks* would look like.

Easykicks Storyboard

Here's one of the simple storyboards that the Peer Insight team started with for *Easykicks*. You'll notice some blank panels; that's intentional as they wanted research participants to co-create this with them. TEMPLATE 4, *page 86,* is a storyboard template you can use in your work.

   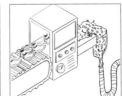 

Assignment

So that's Step 1. Are you ready for your first assignment? Full-scale templates begin on *page 81*.

1

Take the list of ideas or concepts that you have developed and map them onto the **Value/Effort Matrix** [TEMPLATE 2, *page 84*], according to your assessment of the relative value and effort they involve.

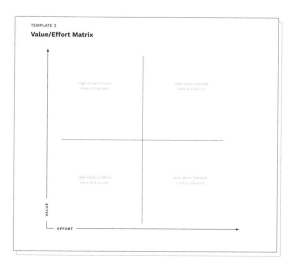

2

Next, consider which concept you want to use to practice the remaining steps in this book and prepare a **Concept Snapshot** [TEMPLATE 3, *page 85*] for it.

In selecting which concept to use, consider the following questions:

- Can you readily access people in the target market you will need to test with?
- Do you have the ability and permission to intervene in their experience?
- Do you have the resources necessary to build a rough prototype, gather data, and run tests?
- Can you engage the support of those needed to help you implement your ultimate solution?

3

Finally, flesh out your concept with a simple **Storyboard** [TEMPLATE 4, *page 86*].

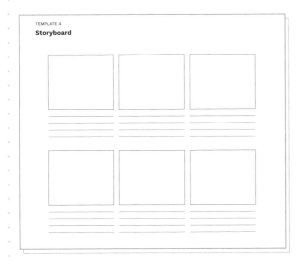

TEMPLATE 4
Storyboard

Great! You've completed Step 1—you've made some choices and now have a concept ready for testing. Log your progress in your **Progress Tracker** [TEMPLATE 1, *page 82*]:

☑ Frame a testable idea

☑ Prioritize concepts for testing

☑ Complete a Concept Snapshot

Now let's move on to the next step in the process—identifying the data you will need to collect to test it.

On to Step 2!

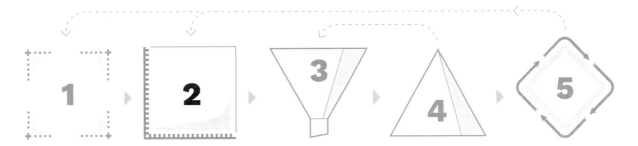

STEP 2

Defining Evidence

IN STEP 2 YOU WILL LEARN TO:

- Surface critical assumptions

- Prioritize assumptions for testing

- Establish what constitutes evidence and identify targets

WITH YOUR PRIORITIZED concept in hand, you are ready to define the kind of evidence you will look for to assess whether your idea is worth further investment. While "fail fast" has become a mantra for innovation efforts, this doesn't mean just tossing ideas into the market to see what happens. Doing quality experimentation requires careful design: If learning is your goal, you want to be sure that your failures are *intelligent* ones that teach you something new and help you make tough choices among competing ideas. That means putting careful forethought into what you are testing for, what success looks like, and what data you need. Here's where we start thinking about what specific data to collect and how to assess whether it supports or rejects our hypothesis. For example, Nike *thought* that people would return their shoes if it were easy for them to do so but wanted to gather evidence to support that assumption.

Definitions

Assumptions	Underlying beliefs about why you think your idea is a good one
Evidence	Countable phenomena or data points that validate or refute a stated assumption—quantitative or qualitative
Hypothesis	The concept you want to test
Threshold target	The minimum value a metric needs to achieve to support moving the concept forward to the next stage of testing
Aspirational target	Future-focused target that describes the value you would like to see the metric achieve when your concept is successfully implemented in the real world
Source	Where you will find your evidence—may be archival (already existing) or new data gathered from the field

To define evidence, we:

1. **Surface the assumptions** behind any given concept;
2. **Prioritize these assumptions** to identify the most critical ones for testing; and
3. **Define the evidence** associated with them and its sources, and **assign target metrics**, both threshold and aspirational. Sources may include both existing archival data and field data (to be gathered in the real world) and can be either qualitative or quantitative.

Sound complicated? It needn't be. We have found that a simple **fill-in-the-blank exercise** can help to identify the different elements in this step.

Let's look at the sequence of activities in more detail. We begin by getting clear about what we are testing for. As we talked about earlier, we are looking for fresh ideas. They need not be new to the world—they just need to add value in three ways: **desirability** (value to our target audience), **feasibility** (ability to be executed), and **viability** (commercial sustainablity over time). We call the intersection of these three conditions the Wow Zone:

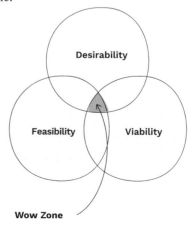

Surfacing Assumptions Fill-in-the-Blank Exercise

Here's an example for SWR's *Concierge* concept:

One critical assumption I have about the

.... is that
concept name

....
assumption

prioritize assumptions

For this assumption to be true, I'd want to

collect ....
quantitative / qualitative evidence

from ....
source

to confirm ....
hypothesis

define the evidence

At the end of this test, we need to see that

.... to know
threshold target

we are on the right track toward our future

aspirational target of ....
aspirational target

assign targets

One critical assumption I have about the *Concierge* concept is that **the concept creates a warm welcome.**

For this assumption to be true, I'd want to collect **comments on the service experience** from **customers who are intercepted for conversations** to confirm **adding *Concierge* at stations will build high-quality personal relationships with customers.**

At the end of this test, we need to see that **at least 50 percent of customers intercepted describe a positive experience with the *Concierge* concept** to know we are on the right track toward our future aspirational target of **80 percent.**

Surfacing Critical Assumptions Underlying Your Concept

To assess whether any given concept wows, we start by surfacing the assumptions we are making about *why* it meets each of the three conditions of desirability, feasibility, and viability. In other words, we articulate clearly why we believe that a concept belongs in the Wow Zone. These assumptions form the foundation of our testing strategy—they will point us toward the right metrics to gather that will signal whether or not these assumptions are valid.

You might be thinking, wait, I'm supposed to test my *assumptions*? I thought I was trying to test the concept itself. But in a hypothesis-driven problem-solving approach, you test the assumptions behind a concept rather than the concept itself. Because our new concepts do not yet exist in the real world (only in our imagination), we don't want to commit the time and resources to build them until we learn more about their desirability, feasibility, and viability. So instead, we test the assumptions underlying their attractiveness. At Whiteriver, they couldn't test the *Electronic Kiosk* without the time and expense of building it—but they could much more easily test the assumption that ER patients were comfortable using computers. This testing of assumptions rather than ideas is how we manage our risk. To test assumptions, we don't need to build a fully functional prototype—we just need a version functional *enough* to test the most critical assumptions underlying it. Without attention to assumptions, we risk investing in flawed ideas, jumping to conclusions,

Assumptions Behind PMI's Spot Concept

Remember *Spot*, one of the four concepts that PMI wanted to test? *Spot* aimed to provide PMI members with real-world experiential learning by matching them with actual, low-risk opportunities to hone their skills. To test *Spot*, the Peer Insight team identified key assumptions under each of the three conditions:

For desirability, they wondered whether members would want to be involved in such projects. Would they see this as a good way to hone their skills?

For feasibility, they wondered if it would be possible to find the right kinds of projects and manage such a complex system of worldwide projects well.

And for viability, they worried that, even if it were feasible to find and run good projects, the cost to do this might be too high to make it viable long-term.

and falling victim to classic cognitive biases like the confirmation bias. Early on, testing the key assumptions underlying a concept, rather than the concept itself, is the fastest, cheapest way to learn.

Surfacing assumptions can turn out to be surprisingly difficult. Counterintuitively, it can be especially difficult for *experts* in different fields. In health care, for example, where the beliefs of highly trained clinical professionals have long driven decision making, assumptions are often made about the needs and desires of patients that turn out not to be true.

False Assumptions in Health Care

IN DALLAS, Children's Health Systems of Texas (CHST), one of the largest pediatric medical centers in the United States, faced clear challenges: the children they served had some of the most troubling health indicators in the United States, with nearly 30 percent living in poverty. CHST leadership realized that their assumptions did not reflect the reality of their patients' lives. They initially assumed that patients and their families focused on preventative care with a proactive mindset, had strong support networks, and trusted caregivers' advice. The reality was that overwhelmed families experiencing poverty often struggled from crisis to crisis with a more reactive symptom-driven focus, lacked traditional support networks, and were sometimes suspicious of caregivers' information. They needed a new approach based on patients' actual experiences, rather than their own beliefs as medical professionals.

IN MELBOURNE, innovators at Monash University Medical Center wanted to train lay telecare guides to act as "professional neighbors" to keep in frequent telephone contact with elderly patients at high risk of hospital admission. They believed that carefully selected laypeople, trained in health literacy and empathy skills and backed by decision support and professional coaches, could reduce hospitalization rates. Many of their clinical colleagues were skeptical, opposed to anyone other than a health professional performing such services, concerned about reducing the quality of care. Rather than debating this point, the innovators engaged their colleagues in the design of an experiment to test the value that lay telecare guides could deliver. Three hundred patients later, the results were in: overwhelmingly positive patient feedback and a demonstrated reduction in hospitalization rates and emergency room visits.

IN ARIZONA, at Whiteriver Hospital, the Performance Improvement Team realized that their assumption that ER patients were comfortable using computers was in error. Many tribal elders, the largest group of emergency room visitors, were not comfortable using the type of new technology that their planned *Electronic Kiosk* employed. Such a system, no matter how efficient it had proven to be in urban Baltimore, would likely create more, not fewer, delays at Whiteriver. As a result, the team pivoted to a paper form aimed at identifying the severity of patients' medical issues as soon as they entered the emergency department. The simple, one-page form literally asked patients whether they needed emergency or nonemergency care, such as a visit with a nurse or a prescription refill, and anyone in the emergency room could help them check the boxes to respond.

As you surface the assumptions behind your concept, you may find yourself tempted to just turn one of the conditions (like desirability) into a statement (Assumption: Users find our solution desirable). But at this abstract level, such an assumption is not testable. We need to push deeper to get more specific about why we believe that this particular condition is met (Assumption: Customers value the time savings that our product provides).

Surfacing assumptions works best as a team sport. Plus, bringing key stakeholders into agreement about what the important assumptions are and what it will take to confirm them is critical. Structured conversations that put the right people in the room (real or virtual) to frame and plan the testing journey and collaboratively identify what success looks like will accelerate your progress by broadening your perspectives and building excitement, momentum, and alignment around what makes a concept a wow. In all four examples we use in this field book, teams worked together to surface assumptions and determine what data they would need to test them.

In the earliest stages of experimentation, assumptions about desirability will be more important than assumptions about feasibility or viability. After all, what good is a feasible or viable solution that isn't desirable to the people it was designed for?

We use the **Surfacing Assumptions** tool (TEMPLATE 5), organized by the three conditions we want to test for, to capture the assumptions related to each of our prioritized concepts.

The team at SouthWestern Railway (SWR) working to test the *Concierge* concept, aimed at building high-quality personal relationships with customers, identified a set of assumptions related to **desirability** (creating a warm welcome that customers valued), **feasibility** (the service can run with existing staff levels), and **viability** (potential to self-fund).

TEMPLATE 5

SWR's *Concierge* Concept Assumptions

Desirability
- Creates a warm welcome that customers value
- Provides more efficient service for customers
- Increases SWR staff enjoyment of job

Feasibility
- Can flex to the needs of different stations
- Allows skills to be developed without costly training
- Can be run with existing staff
- Employs existing assets

Viability
- Allows technology to be integrated to achieve self-service
- Frees up potential increased retail space
- Makes self-funding possible

When you lay out the assumptions underlying your concepts, you will likely find that you have more than you can—or want to—test. Beware of letting the scope and cost of experiments balloon by trying to test too many assumptions. In Step 1, we prioritized our concepts to determine which ones to move into testing. Now, in Step 2, we will dig deeper and prioritize the critical assumptions underlying each concept. That will tell us where to start the testing process. Fortunately, not all assumptions are equally important—you start with the most critical ones, the ones that make or break your concept.

Prioritizing Assumptions for Testing

Almost all new concepts rest on a limited number of particularly significant assumptions. If these are true, they make the concept worth moving forward. If they are false, the rest of the assumptions don't matter enough to bother testing. To prioritize, focus your attention on two aspects of each of your assumptions:

1) How critical is that assumption to the success of the concept?

2) How much do you already know (from existing sources) about the likelihood that this assumption is true or false?

You can use the **Prioritizing Assumptions** tool (TEMPLATE 6) to capture this.

After putting together their list of assumptions for the *Concierge* concept, the SWR team prioritized their critical assumptions before proceeding.

Prioritizing Critical Assumptions Behind the *Concierge* Concept

The DK&A team sorted the assumptions behind SWR's *Concierge* Concept, using the criteria of how important each was to the concept's success and what they already knew. Some, such as the possibility of self-funding and the use of existing assets, were considered to be low priorities for testing: they were not seen as critical, and quite a lot was known about them. Others, such as the training required to develop the new staff skills needed and the ability to integrate new technology, were seen as more critical, but SWR had confidence in their existing knowledge of them.

Three assumptions emerged as the most critical and unknown, all related to desirability.

The *Concierge* concept:
1. Creates a warm welcome that customers value
2. Provides more efficient service
3. Increases SWR staff enjoyment of their jobs.

These three assumptions formed the focus of their initial testing efforts.

Translating Assumptions Into Evidence

Once you have identified a small number of critical assumptions to prioritize for testing, we move from talking about what we *assume* to be true to talking about collecting *evidence* that it is true. This transition changes the conversation in two significant ways. First, it becomes personal. Assumptions reflect properties of the new concept and should be visible to all—but whether something is "proven" true lies in the eyes of the beholder. Pay attention to who needs to conclude that something is true and how they see the world. We tailor the identification of metrics to the key stakeholders we must convince. Invest some time in thinking about who those stakeholders are and what their relationship is to the concept you want to test. Who needs convincing? How skeptical are they? What is at stake for them? How fast do they expect you to generate evidence?

You can use these questions as an exercise in co-creation—invite a diverse set of your key stakeholders

⚠ **Warning**

You may be tempted to skip the surfacing of assumptions and go straight to identifying evidence, based on the data you have available. This is a serious mistake, as it will encourage you to work backward from the data you've got. That is not hypothesis driven! Instead, you want to start with the assumptions you need to test and only then specify the right kind of evidence to assess them (which may or may not already exist).

into a conversation aimed at answering them. At SWR, for example, one of the key stakeholder groups, senior SWR leadership, was anxious to see credible test results in actual station environments as soon as possible.

The second change in the conversation, as we move toward collecting evidence is the level of specificity. Though we have worked hard to make our assumptions less ambiguous, evidence must be observable and countable, a more demanding standard. It must also fit the context of the test you are about to do. Some issues to consider as you define your evidence:

- How much time do you have available for the testing process?
- What kinds of resources can you call on?
- How big is your budget? What can you afford?
- What is possible from a technical viewpoint?

The DK&A team asked themselves how they would know if the three critical assumptions behind the *Concierge* concept were true. How would these show up in observable and countable ways? To test the **creates a warm welcome** assumption, they considered evidence based on how many customers sought out the *Concierge* services and whether they expressed appreciation for the services. For the **increased efficiency** assumption, they looked for reduced wait times at windows and increased use of self-service. For **improved employee satisfaction**, they solicited staff members' feelings about their new role.

The form that evidence takes often needs to be iterated as you push further into the experimentation process and learn more. In general, as projects progress, the desired evidence increases in terms of specificity, becomes more quantitative, and comes from multiple sources to manage the risk of the increasing investment being made in the new concept. In the early stages of testing, it is not always obvious what to measure. Data sources in the real world are messy, and more than one metric is often needed to show whether assumptions are true or not. Sometimes the best we can do is to verify whether we are *directionally* correct. For these reasons, triangulation from multiple data sources is always valuable.

It makes sense to begin your search for the right evidence by reviewing those measures already in use that may have value and asking yourself the following questions: What evidence has meaning in this sector, with the audience you must convince, or for this type of activity? What is already being measured? Can you adapt ongoing monitoring to support your experiment? Examining existing data is an important step in preparing to go into the field—just make sure to remain hypothesis driven!

The relative value of *qualitative* versus *quantitative* data is also important to consider. Quantitative data may be already available, but does it measure what you need? Qualitative data, the kind that is good for sense-making and that assures directional correctness, may be essential, but needs to be gathered from the field. In addition, some audiences may prefer

TEMPLATE 7

Assumptions to Evidence for SWR's *Concierge* Concept

Assumption	Evidence	Threshold Targets	Source
Creates a warm welcome that customers value	Frequency of customers seeking interactions with *Concierge* staff	Number of customers interacting with *Concierge* staff goes up 30 percent or more	Observation in ticket hall
	Positive comments from customers expressing appreciation for the service	60 percent of interviews have customers noting appreciation for the service when unprompted	Customer intercept interviews
Provides more efficient service for customers and enables improved interaction with SWR staff	Volume/speed/accuracy of *Concierge* staff answers	90 percent of *Concierge* staff answers meet customers needs in terms of volume/speed/accuracy	Observation in ticket hall
	Length of queues at ticket windows	40 percent (or more) decrease in length of queues	Observation in ticket hall compared with historical record
	Increased volume of tickets purchased at self-service window	40 percent (or more) increase in volume of tickets purchased via self-service machines	Sales data from self-service machines compared with historical record
Increases SWR staff enjoyment of jobs	Staff appreciation for and interest in remaining in new role versus returning to ticket window	60 percent of staff have an appreciation for and express interest in remaining in new role	Staff interviews

quantitative over qualitative data, while others love stories. As you specify the evidence for your concepts, you will likely want a combination of qualitative and quantitative data, drawn from both archival and field sources, at each stage. Once these are specified, you need to identify the source of the data. Ask yourself *where* you will find the data you need.

The DK&A team, for instance, identified evidence in support of the *Concierge* concept that was both qualitative (e.g., customer and staff satisfaction) and quantitative (e.g., queues at ticket windows, sales data from automated ticket machines).

As a final activity in Step 2, it is valuable to identify and differentiate between aspirational targets (the ones that you hope your idea will eventually produce) and threshold targets (which tell you whether to go to the next step) and offer both for each metric. It is unrealistic to believe that you will reach aspirational targets over the course of hours, days, or weeks. Even so, setting a target up front provides information to assess the magnitude of change and how long it might take to reach the aspirational targets. You can then consider whether that aligns with your project timeline to reach the desired impact or not. If yes, carry on. If not, you may need to go back to Step 1 and select a different concept for testing.

Tips for Surfacing Assumptions

- **What questions are still outstanding?** Ask yourself, "What don't I know for certain?"

- **Put on an investor's hat:** What questions would need to be answered to receive another round of funding? What do you need to prove?

- **Convert your questions into affirmative statements** (e.g., Customers want to recycle old shoes).

- **Think about the categories:** desirability, feasibility, viability.

- **Be sure your assumption is specific enough.**

- **Revisit some of the design tools** traditionally used to inspire idea generation, like journey maps, jobs-to-be-done, personas, or value chain maps, to help you surface assumptions.

Tips for Identifying Targets

For each piece of quantitative **evidence, ask yourself:** What is the smallest amount of change you'd want to see in order to feel solid about moving this concept forward for further testing? This will give you your threshold target. Then ask: What is the desired/hoped for amount of change when the concept is successfully implemented? This helps pinpoint your aspirational target.

For each piece of qualitative **evidence, ask yourself:** What are the near-term expected responses that would signal that you are headed in the right direction? These will give you a threshold target. Then ask: What are the desired/hoped for responses you would want to see in the future? This signifies your aspirational target.

Assignment

It's assignment time again! (Remember that you can find the full-scale templates beginning on *page 82*.)

1

Use the **Surfacing Assumptions** tool [TEMPLATE 5, *page 87*] to lay out the key assumptions you believe underlie the concept you have selected for testing, according to the conditions of desirability, feasibility, and viability.

2

Next, identify which assumptions are most critical, and decide which ones it is most important to test first using the **Prioritizing Assumptions** tool [TEMPLATE 6, *page 88*].

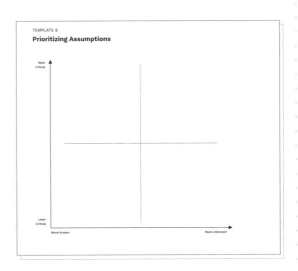

TEMPLATE 5
Surfacing Assumptions

Concept Name

Desirability

Feasibility

Viability

TEMPLATE 6
Prioritizing Assumptions

Most Critical

Least Critical

More Known   More Unknown

3

Finally, using the **Assumptions to Evidence** tool [TEMPLATE 7, *page 89*], capture the evidence, sources, and targets associated with each assumption.

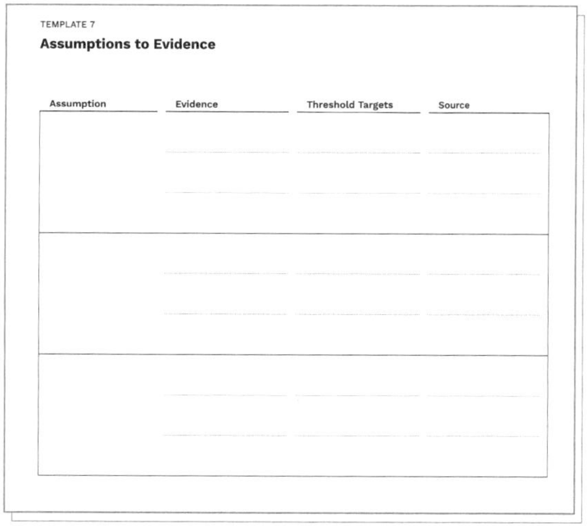

ASSIGNMENT

TEMPLATE 7
Assumptions to Evidence

Assumption	Evidence	Threshold Targets	Source

Log your progress in your **Progress Tracker** [TEMPLATE 1, *page 82*]:

☑ Surface critical assumptions

☑ Prioritize assumptions for testing

☑ Establish what constitutes evidence and identify targets

Now that you have a clear set of assumptions, metrics, and targets for the concept you want to test, you are ready to move from the *what* of testing to the *how*. Then move on to the next step in the process—selecting a type of test.

On to Step 3!

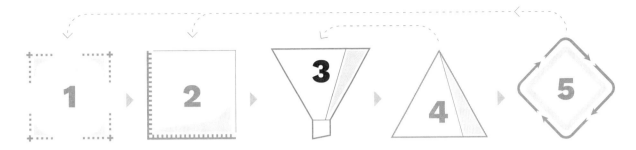

Selecting Your Test

IN STEP 3 YOU WILL LEARN TO:

- Choose between Say and Do tests

- Select a test type

- Design your test

IN STEP 2, you identified the critical assumptions underlying the concept you want to test, along with associated evidence and targets. Now it's time to select the type of test that will help you gather the data you need to assess whether your critical assumptions are valid.

Keep in mind that you will run multiple tests over the course of the experimentation process, which will grow in complexity and rigor as your concept evolves. Nike didn't run just one test on *Easykicks*, their shoe subscription service. Instead, they started with one small test, and when the data showed that they should keep going, they ran further rounds—a second test, and a third, and so on. Here, in Step 3, we will lay out a set of questions that will allow you to select the best test at each particular point in your journey.

There are many different kinds of tests and selecting the best one at any given point in your learning

Five Types of Field Tests

Cognitive walkthroughs help you find partners and customers by walking them through your concept in detail. They invite the prospective user to help you shape a still malleable concept, using tools like pitch decks, storyboards, or paper prototypes.

Lemonade stands are short-term pop-ups, easy to create and then easy to take down. Though they represent a more significant investment than cognitive walkthroughs, they allow the user to come to you, rather than you going to them.

Smoke tests run competing versions of ads with different value propositions (how we propose to deliver value) linked to landing pages aimed at "faking a new business fast." They ask prospective customers to act in some way—click a link, send an email, or sign up to be part of a beta test.

Simulations can be either virtual experiences that prospective users click through or physical experiences that use a Wizard of Oz "man behind the curtain" delivery approach. Both focus on providing an authentic front end for users to experience but lack functioning back ends.

Trials represent the most elaborate type of test design and use fully functional prototypes, with authentic front and back ends. They often rely on accompanying methods like control groups and pretest/post-test designs to ensure more rigorous measurement of outcomes.

process can seem daunting. This chapter will guide you through a series of questions to sort through your options and help you determine the best one. We know that there are a lot of concepts, assumptions, evidence, and questions swirling around in your brain right now. That's to be expected. We will take all of these inputs and systematically lead you to the single test type that will best help you learn based on where you are right now. This chapter is a funnel of sorts, guiding you to take vast amounts of information and determine a way forward.

To simplify this, we will focus on just **five types of field tests**—the ones we observe to be used most often in the experimentation process. ("Field test" means a test run in the real world, or in the "field.")

Choosing the type of test that best suits your circumstances and stage of experimentation requires stepping back to consider a set of fundamental questions about the concept to be tested, your goals, the environment you are operating in, and those you are designing for. As your concept moves through cycles of testing, you will run multiple tests. So, you will revisit this step again and again, repeating these fundamental questions at each stage or round, refining your plan as your concept evolves.

Keep in mind that no test will allow you to conclusively "prove" that your concept is a good one. Until you build and pilot it in the real world, you cannot know for sure. Your goal in early tests is to better manage the risks of investing in any given new concept by assessing whether you are directionally correct enough, relative to your critical assumptions, to move to the next stage of experimentation.

In general, the tests involved in a concept's evolution change over time from simple, fast, and inexpensive (like cognitive walkthroughs) to more complex, lengthier, and expensive (like simulations and trials). Selecting an appropriate test has a lot to do with where your concept is in its development and situational context. By asking yourself a series of questions, you can focus your search for the right test to do at any point in the experimentation process.

Let's envision your journey to select a test type as a **flow of decisions** that asks four questions at different decision points, in a particular order. After presenting the entire decision flow, we will break it down question by question.

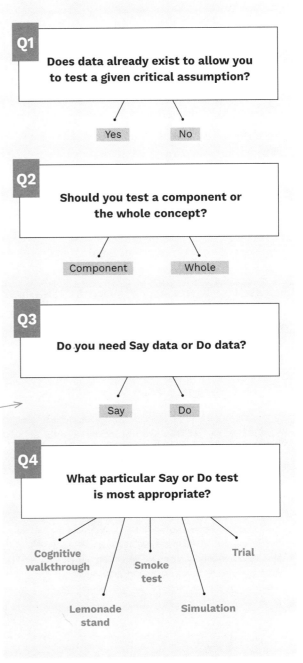

Test Design Decision Flow

Q1 Does data already exist to allow you to test a given critical assumption?

Yes   No

Q2 Should you test a component or the whole concept?

Component   Whole

Q3 Do you need Say data or Do data?

Say   Do

Q4 What particular Say or Do test is most appropriate?

Cognitive walkthrough   Smoke test   Trial

Lemonade stand   Simulation

Question 1

> **Does data already exist to allow you to test a given critical assumption?**

The answer to this question will determine whether a thought test or a field test makes sense to start with. Though the heart of experimentation lies with field tests that gather data in the real world, you don't want to ignore any existing archival or secondary data that could save time and money and get you off to a quicker start in testing your assumptions. Existing data can often be cheaper and faster to obtain and may have higher credibility with your key stakeholders. Of course, your new concept can never be tested directly with data from the past (it didn't yet exist!), but you may be able to use existing data to test particular critical

Are Thought Tests Really Tests?

Some might argue that thought tests aren't really tests at all—they look like traditional analysis. It is true that they lack the interactive component that our five types of field tests have. We like the term *thought tests* because it reminds us to be hypothesis driven. Most of us have been taught to take the data we've got and work backward from it to answer important questions about our new ideas. Hypothesis-driven thinkers start with the concept or assumption they want to test and, whether in the field or using existing data, begin by asking what data they need to test it. By switching the logic, we transform a traditional analysis into something more rigorous and experimental.

Don't Forget to Have Fun!

We don't want to lose the spirit of experimentation and make the process sound onerous; tests should be fast, energetic, and fun. The very act of running the test drives commitment and momentum because it builds confidence through evidence. It's exciting to see the data rolling in that offers real evidence to either confirm or disconfirm key assumptions underlying your concept. But taking the extra time to design your test thoughtfully ensures that your ultimate test outcomes will be about the quality of your concept—not the result of flaws in your testing choices.

assumptions. We call these kinds of tests *thought* tests. Remember your friend who had the idea for a drone book light for reading at night? They might first run a thought test that uses existing secondary data, finding out the percentage of people who read paper-based materials at night in bed and the percentage of those who use a reading light. How big a market does that translate to? Trend data would also be useful. If there is a downward trend in sales of book lights (maybe from people using electronic devices to read?), this would be helpful to know before developing the idea.

Thought tests start with a specific assumption you want to test and ask what you already know—or might easily learn—that would offer evidence for whether it is true or not. So, the first question to ask in Step 3 is whether existing archival data might shed light on any of your critical, make-or-break assumptions.

It is likely that relevant data exist for some of your critical assumptions, either already *known* to you or *knowable* with some effort. Most likely, you will also have critical assumptions—often relating to key stakeholder behaviors—that are *unknowable*: good data simply won't exist until you move into testing in the real world. The **Data Sort** tool (TEMPLATE 8 at the end of this step) can act as a guide as you sort through your make-or-break assumptions, using these three categories: what you already *know* today (based on existing data), what might be *knowable* with some investigating of data sources, and what is *unknowable* until you collect new data in the field.

For example, at SWR, DK&A ran a thought test on the larger architectural and real estate project that the *Concierge* concept rested upon. They calculated that if they could redesign the entire ticket hall experience, eliminating most of the ticket booths, space could be freed up for additional value-added retail. Knowing the approximate cost of the full architectural design and rebuilding project, they were able to test the financial viability of the idea using readily available estimates for retail revenue generated per square foot. On this basis, they determined that the idea had promise without field testing it.

Remember Whiteriver Hospital's simple form that asked patients as they entered the emergency room whether they needed urgent care or not? The legality of that approach was obviously a key assumption to be tested. The answer was unknown to the team but clearly *knowable*. HHS staff connected them with the Centers for Medicare and Medicaid Services, who informed them that the U.S. Emergency Medical Treatment and Active Labor Act of 1986 made the use of any pre-examination form illegal. An effort to prevent emergency rooms from turning away those without insurance, this "no dumping" law stipulated that everyone who came into an emergency department must be assessed by a medical clinician. Though their concept was intended to help, not send away, ER visitors, the team's assumption of legality had been disconfirmed, necessitating another pivot to a new concept. Abandoning the paper form, they circled back and devised a new concept: a fast-track system that placed medically qualified personnel at the emergency room entrance to quickly—within fifteen minutes—assess each patient's condition and direct visitors to appropriate nonemergency services.

To test critical assumptions when existing data sources are insufficient, you must move into the field to answer a new set of questions.

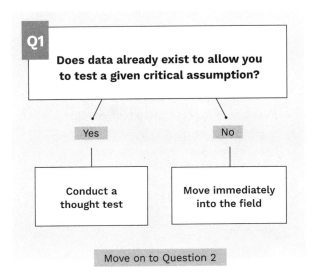

Question 2

> **Should you test a component or the whole concept?**

Next, we ask whether we should test specific *components* of a concept or the *whole* concept. To answer this question, consider the following:

- Do the most critical assumptions to be tested belong to specific elements of the concept, or do they belong to the concept as whole? Answering this will help you determine if you want to test a key component or the entire concept. It is often possible to do either—most concepts consist of a set of elements or subsystems that can be either decoupled for testing or tested together as a whole. Nike, for instance, could decouple parts of the *Easykicks* concept, separating the subscription and recycling ideas, or test the whole concept. If critical assumptions relate to a specific component of the concept, it is usually faster and more appropriate to test that component first. Critical assumptions are best tested sequentially, or at least in parallel experiments, not all together.

In other circumstances, testing components individually may be problematic. Consider these questions:

- How difficult is the full concept to test? Some concepts, particularly ones that are novel, can be expensive and time consuming to test. For these, component testing can be an especially effective way to reduce risk and investment during early tests. Other concepts are relatively easy to test as a whole, and even though assumptions are tested simultaneously, they can still be individually assessed. Delaying the testing of the entire concept in order to test components sequentially may not make sense.

Testing Components of the *Easykicks* Concept

Peer Insight's *Easykicks* team felt that it made sense to test individual components of the concept separately, rather than the whole concept, for several reasons. First, the critical assumptions they surfaced belonged to specific components, such as assumptions about customers' willingness to return old shoes. In addition, testing the whole *Easykicks* concept would have required significant upfront investment that Nike wanted to avoid early in the experimentation process. They were also concerned about signaling to competitors that they had a new idea. So, Peer Insight opted to start with a focus on the shoe recycling component alone, testing the assumption that parents would want to recycle used sneakers and would actually mail them back to the company.

- How sizable are the synergistic effects of all the components working together? In some concepts, the synergistic "gestalt" effect (where the total is more than the sum of its individual parts) is an important dimension—components' interactions create additional value. Where this is true, you risk a false negative if one component, tested separately, fails. The concept, tested as a whole, might still have succeeded. If you anticipate the synergies to be strong, you may need to test the concept as a whole, with the components working together. This requires extra diligence in planning the test so that you can isolate the effects relative to the individual assumptions.

In making your choice about whether to test a component or the whole, consider the balance in your answers to these qualifiers:

Q2

Should you test a component or the whole concept?

if

☐ Critical assumptions belong to specific elements

☐ Trialing the whole concept is expensive and complex

☐ Synergies are not significant

Test critical components first

if

☐ Critical assumptions belong to the whole concept

☐ Trialing the whole concept is not more difficult than trialing components

☐ Synergies are significant

Test entire concept

Move on to Question 3

Testing the *Concierge* Concept as a Whole

DK&A and SWR took the opposite approach from the *Easykicks* team, deciding that it made sense to test the *Concierge* concept as a whole. Synergies among the components were significant, they believed, and testing the whole was not substantially more difficult than trying to break its components apart and test them individually. DK&A was confident that they could still effectively test individual assumptions (like creating a warm welcome versus increasing staff satisfaction). Furthermore, timelines were short, and SWR leaders were anxious for field data as soon as possible.

Question 3

┌───┐
│   │
│   **Do you need Say data or Do data?**   │
│   │
└───┘

Having decided whether to test a single component or the full concept, we now consider how best to test it. One of the significant differentiators of experimental approaches is whether they produce Say or Do data. Say data comes from tests that *ask* prospective users what they will do or what they like or need; Do data is based on their *actual behaviors*. In reality, the different types of tests fall along a continuum. At one extreme, in pure Say approaches, we often seek out prospective users and bring the concept to them. We then help them navigate through the experience. In Do approaches, we wait for them to come to us and leave them to navigate the experience without our intervention. The gold standard for evidence is Do data—a tidal wave of research demonstrates that people frequently *say* one thing and *do* another. Ultimately, Do data will be required to establish proof, but such data can often be expensive and more time-consuming to get, and may require a more complete prototype (more on this in Step 4). Early in the testing of a new idea, to limit risk and learn quickly, Say data is usually the preferred choice, with the addition of simple prototypes to improve the accuracy of feedback.

How do you decide? Again, another set of questions can help assess whether Say or Do data will make the most sense at any given point in your experimental process. Some of these questions relate to you, the innovator, your goals, and the context you are working in. Others relate to the concept you are testing, and still others to the behavior of potential users. As you decide the relative value of Say versus Do approaches, consider the following:

- What are your goals for the research? Along a continuum, your research goals may be generative (focused on better understanding the needs of users) or evaluative (focused instead on testing a specific solution). Generative goals mesh well with Say approaches that invite prospective users to share their own ideas and motivations or "think aloud." Do approaches provide more accurate data on actual behavior and allow you to test aspects of the user experience, like awareness, acquisition, and retention, that Say approaches cannot test reliably.

- What are your key stakeholders willing—or even able—to invest in terms of time and resources in testing? What is their standard for proof and their appetite for risk taking? These closely related factors often work at cross purposes. Innovators often work with stakeholders who are unwilling—or unable due to other pressures or priorities—to invest significant time or money in testing. These constraints can push them toward faster, more economical Say testing. However, other stakeholders (or sadly, sometimes the same ones) have a high standard for proof, only achievable with Do testing. In another vein, some stakeholders have a high tolerance for failure and want to take visible, bold actions, favoring more dramatic Do approaches. Others shrink from any visible failures, necessitating the kind of "under the radar" testing that favors the simplicity of Say tests.

- What developmental stage is the concept at? Are you open to pivoting or unwilling to make significant changes? Allowing prospective users to shape a concept, using the kind of inclusive methods characteristic of Say approaches, can be very powerful—but requires a willingness on the innovator's part to make substantial changes in the concept. Once a new idea's development has evolved to the point where it is "baked" (in the mind of the innovator or key stakeholders), it makes sense to progress to the kinds of Do tests that provide more specific evidence of a concept's success or failure under real-world conditions.

- How much change does your new concept ask of users? Can it be easily adapted to their current routines, or does it require significant changes in their behavior? Say approaches rely on a prospective user's ability to imagine a new experience and predict their own future behavior. The more radical the behavioral change a new concept requires from users to adopt it successfully, the more likely their imaginations will fail and only Do approaches will provide reliable data about their actual response to the concept in practice.

So, consider the following qualifiers:

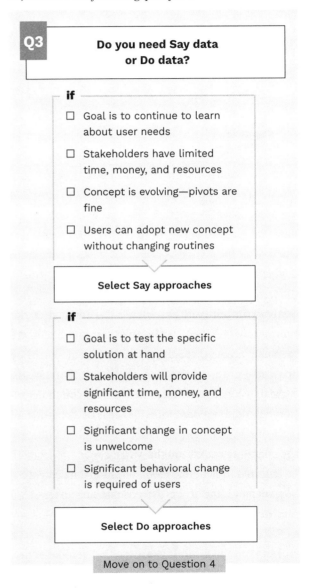

The **Say/Do continuum** (TEMPLATE 9 at the end of the chapter) will help you assess the strengths of the specific test types.

Question 4

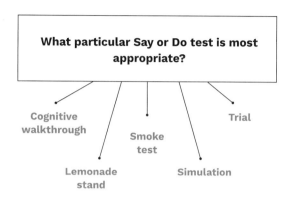

Based on your answers to the three previous questions, let's look at the strengths of each of these five types of field tests (refer back to page 34).

Cognitive walkthroughs anchor the Say end of the continuum. They help you find partners and customers by walking them through your concept in detail. They invite the prospective user to help you shape a still malleable concept, using prototypes like pitch decks, storyboards, or posters (more on these in Step 4).

Use cognitive walkthroughs when:

- Your goal is generative—you want to deepen your understanding of users' needs rather than test a specific solution.
- Say data is acceptable.
- You want the flexibility to test either components or the concept as a whole.
- Learning quickly and cheaply is critical.

"Must-have" ingredients for successful cognitive walkthroughs include a clear value proposition that you want to test (i.e., how will it meet the needs of the users you are designing for?) and simple paper-based prototypes that capture that value proposition. These are not written business plans; they are something that the test users can interact with. In the B2B world (businesses or organizations selling to or serving other businesses or organizations), prototypes for cognitive walkthroughs often take the form of pitch decks and are shorter and less in-depth.

Cognitive Walkthroughs at SWR

The DK&A team began testing the three high-priority concepts at SWR (the *Concierge*, *Clean and Cleaner*, and the *Wayfinding Audit*) using cognitive walkthroughs with the relevant station manager where each concept was to be tested. These interviews were quick to arrange and inexpensive to conduct. Even more important, the need for the station managers' support for whatever solution they pursued made them key stakeholders to engage, and each had a clear commitment to customer service that made them a good proxy for their customers. Though the DK&A team had already developed an initial version of the concepts, they were anxious for feedback, positive or negative, and open to iteration. They presented a detailed briefing on the *Concierge* concept to Basingstoke's station manager, for example, elaborating on the concept's vision and its components. Based on station managers' feedback, the team made further iterations and created a refined set of concepts they felt were ready for the next level of experimentation, moving from Say to Do data.

Lemonade stands, or pop-ups, represent a more significant investment than cognitive walkthroughs. They allow the user to come to you, rather than you going to them.

Use a lemonade stand when:

- Your prospective users gather at some venue where you can locate them en masse, such as an industry conference, membership event, or retail store. You can also opt to create a special event, as the **City of Dublin** did.
- Your goals are evolving from generative toward evaluative—you want to test a specific end-to-end solution but are still open to pivoting.
- Say data alone is not enough, and you want to begin transitioning to Do to see if prospective users are drawn to a concept. Yet you are still exploring and want to be able to interview them to better understand their motivations and actions.
- You have sufficient time and money to create appropriate props (in the form of a booth or end-of-aisle display) and prototypes, and staff the stand.

"Must-have" ingredients for lemonade stands include a venue that offers the opportunity to attract the attention of potential users. You also need some sort of "booth" (sometimes just a simple table) to display prototypes (often posters). You can include the option to indicate interest in one of the prototypes by offering potential users a flyer that has a QR code, providing some preliminary Do data. Potential users can also be "intercepted" and interviewed. A tight plan for data capture is especially critical, as there may be multiple

Is this an experiment or a fair?

Experiments don't have to be dull—and lemonade stands can take a variety of forms. In Dublin, Ireland, citizen innovators hosted a one-day event that they called a "Prototype Extravaganza" to attract as many residents as possible and get their feedback on a set of new ideas the group had created. In a kind of market fair atmosphere, attendees visited a series of booths that presented the different concepts. The energy of the extravaganza helped secure champions for the ideas that moved forward.

Utilizing a Lemonade Stand at PMI

The Peer Insight team began the transition from more exploratory conversations around the four potential PMI concepts to more evaluative ones, from purely Say data to an element of Do. They took advantage of a large conference where a significant number of PMI members were gathered. They designed a lemonade stand in the form of a prominent booth that contained posters and handouts illustrating the value proposition of each concept being tested. Members drawn to the booth were invited to "enroll" in any concept that interested them. The Peer Insight team observed positive and negative reactions and captured whether people said they already used analogous services. Members signed up to learn more through an enrollment form that collected valuable information to help the team deepen their understanding of member needs and profiles. Accompanying intercept interviews provided deeper insights that helped the team further refine and prioritize the four concepts.

potential users approaching the stand at the same time (unlike the slower pace of cognitive walkthroughs).

Smoke tests derive their name from the testing of electronic equipment and the dictum "if it smokes when you plug it in, unplug it." You can also think of it in terms of the well-known "smoke and mirrors" metaphor—you're setting up a fake test that isn't real. Smoke tests take the form of ad tests (running competing versions with different value propositions to see what gets clicked) and landing pages that allow you to "fake a new business fast." They ask prospective customers to act in some way—click a link, send an email, or sign up to be part of a beta test.

Use a smoke test when:

- The component of the concept that you are most interested in testing is the value proposition.
- Your research goals are moving toward the evaluative end of the continuum, and you have specific value propositions to test.
- You want to gather arms-length Do data on interest from prospective users and prioritize their needs.
- You have sufficient money to purchase ads and time to run an ongoing test.

"Must-have" ingredients for smoke tests include a detailed understanding of what you are trying to learn, as well as expertise in writing and managing online ads. These tests also require the use of a platform (e.g., Google or Instagram) that is appropriate for testing. They benefit from clear plans for data capture and

Using Smoke Tests at PMI

The PMI team also used smoke tests to further refine their understanding of the relative merits of the four concepts. Using LinkedIn ads and emails, they presented a variety of value propositions to see which succeeded at gaining traction among members. Potential customers then went to a landing page that further articulated each value proposition and its features. The team gathered data on email open rates, site visits, and the percentage of site visitors who gave email addresses and who downloaded resources or watched a video. A subset of these enrollees was interviewed about their choices.

analysis, including a list of what the test won't provide. Prototypes, while of lower fidelity than those used in trials or simulations (e.g., simple sketches rather than almost-finished physical or digital versions of the product or service), should still appear "real" to the end user. Landing pages should be visually appealing, potentially with simple functionality like inviting users to provide their email address to learn more.

Simulations can be virtual experiences that prospective users click through or physical experiences that use a Wizard of Oz "man behind the curtain" delivery approach. Both lack even a minimally functional back end.

Use simulations when:

- Your goal is evaluative, and you want to test a specific solution. (If you want to add a more generative dimension, prospective users can be asked to "talk aloud" as they navigate the simulation.)
- Gathering Do data is important to you.
- You are interested in testing a full end-to-end concept.
- Saving the time and/or money required to build even a minimally functional back end is attractive.

Creating a Simulation for an *Easykicks* Component

The *Easykicks* team decided to move from exploratory conversations to evaluative ones and created a simulation as part of that process. Still electing to test a single component—the recycling of shoes—they wanted to test the assumptions that people were willing to return old shoes and that they valued this option. To do so, they created a fully authentic front end in which people buying shoes were provided with a stamped addressed bag and invited to recycle their old shoes by mailing them back. Peer Insight employees handled the returns at their office, "faking" a back-end operation using a "Wizard of Oz" approach.

Simulations can be thought of as higher-fidelity cognitive walkthroughs, because in both cases, users know they are walking through an experience that is part of a research plan. They can take the form of a one-on-one research session or be part of a diary study/session where users are alone and are asked to complete tasks and talk aloud. Digital movements through the prototype can be tracked with the appropriate software. Users are put into an overall use case, but in a more open-ended context.

"Must-have" ingredients for simulations include a strong data capture plan that aligns with the critical assumptions you're trying to collect data to examine. You'll be synthesizing qualitative data from users (likely talking out loud as they use the prototype), but you might also ask them for qualitative feedback afterward. This feedback will need to be analyzed alongside data about their actual behavior (captured through software or observation). The other "must-have" in your research plan is a way to capture the "why" behind the data. This is largely something you can learn from qualitative research, so try to include a direct interview with the research participant (probably after they interact with the prototype) for their "voiceover."

Trials represent the most elaborate test type and use functional prototypes. They often rely on accompanying methods like control groups and pretest/post-test designs to ensure more rigorous measurement of outcomes. This is the land of the minimum viable product (MVP) prototype.

Use a trial when:

- Your research goal is evaluative (though you can add additional methods like diary studies and interviews to provide deeper generative data as well).
- You have confidence in your understanding of users' needs; the desirability box is checked, and now you want confidence in viability.
- Key stakeholders expect Do data and are willing to provide sufficient time and money to obtain it.
- It is possible or even preferable to build a functional back end.

Successful trials begin with a cross-functional team ready to respond quickly, to react to user feedback, or to pivot. Trials usually need to run over a longer period. Planning, in the form of follow-up interview or diary study questions or scenarios, key learning objectives, flows, roles of interviewers, and/or prompts, is key. Compensation for users' time and feedback is recommended and sometimes necessary. Trials in this context are not synonymous with randomized controlled trials (RCTs), which can last years and are done with high levels of rigor. The experimentation process described in this field book should happen prior to dedicating resources, time, and effort to RCTs.

Designing a Trial for SWR's *Concierge* Concept

The DK&A team was anxious to test the *Concierge* concept in an actual station environment. Would it, in fact, create a warm welcome for customers, help them better navigate their journey, reassure them that train usage was safe, and encourage train use for leisure? Would staff be able to better interact with and support customers? Would it increase the enjoyment and sense of pride that the staff felt? They opted to undertake a trial of the entire concept—with a real front end and back end—at Basingstoke Station over a five-day period. DK&A changed the physical environment of the station, trained staff in new roles, and observed how customers experienced the new concept. In addition, they did customer intercept interviews in real time and conducted staff interviews afterward.

Let's summarize the factors involved in answering Question 4:

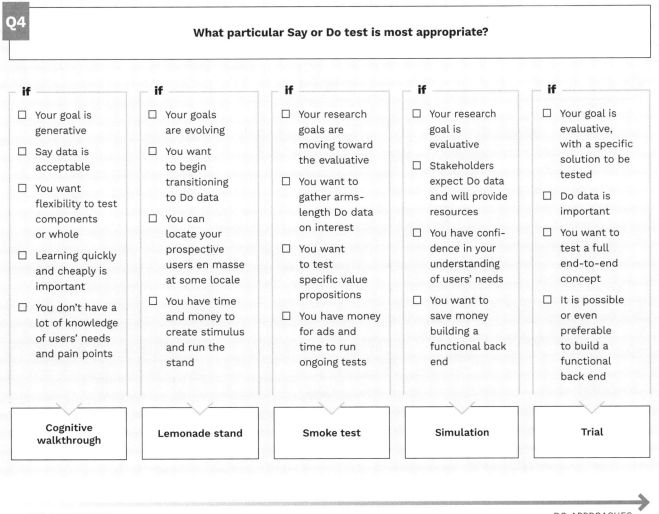

Q4   **What particular Say or Do test is most appropriate?**

if	if	if	if	if
☐ Your goal is generative	☐ Your goals are evolving	☐ Your research goals are moving toward the evaluative	☐ Your research goal is evaluative	☐ Your goal is evaluative, with a specific solution to be tested
☐ Say data is acceptable	☐ You want to begin transitioning to Do data	☐ You want to gather arms-length Do data on interest	☐ Stakeholders expect Do data and will provide resources	☐ Do data is important
☐ You want flexibility to test components or whole	☐ You can locate your prospective users en masse at some locale	☐ You want to test specific value propositions	☐ You have confidence in your understanding of users' needs	☐ You want to test a full end-to-end concept
☐ Learning quickly and cheaply is important	☐ You have time and money to create stimulus and run the stand	☐ You have money for ads and time to run ongoing tests	☐ You want to save money building a functional back end	☐ It is possible or even preferable to build a functional back end
☐ You don't have a lot of knowledge of users' needs and pain points				

Cognitive walkthrough	Lemonade stand	Smoke test	Simulation	Trial

SAY APPROACHES → DO APPROACHES

Assignment

Now that we have covered the entire decision flow, it's your turn again!

1

Start with the selected concept from Step 1, for which you have identified critical assumptions and defined evidence and targets in Step 2. First, use the **Data Sort** tool [TEMPLATE 8, *page 90*] to evaluate the usefulness of the existing data available to you.

2

Next, as you prepare to move into the field, use the **Say/Do Continuum** [TEMPLATE 9, *page 91*] to decide whether Say data will be best to test the assumption or Do data will be important.

TEMPLATE 8
Data Sort

Assumption	EVIDENCE		
	What We Know	Knowable with Existing Data	Knowable Only with Field Data

TEMPLATE 9
Say/Do Continuum

Put an X or a dot along the continuum noting the thoughts of your team and stakeholders. The farther your dots are to the right, the more important Do data will be.

Generative	Evaluative
Research designed to generate possibilities	Research designed to see if a solution is working

Beer budget	Champagne budget
Not much time, money, and resources	Ample time, money, and resources

Willingness to pivot	No major pivots
Open to user feedback that could fundamentally shift concept	Focus is on improving the main idea; concept is mostly "baked" already

Easy to adopt	New experience
Concept flows easily into current routines and behaviors	Concept requires significant change in behavior and norms

3

Now you are ready to use the **Test Design Decision Flow** [TEMPLATE 10, *page 92*] to select the most appropriate test for your concept and situation.

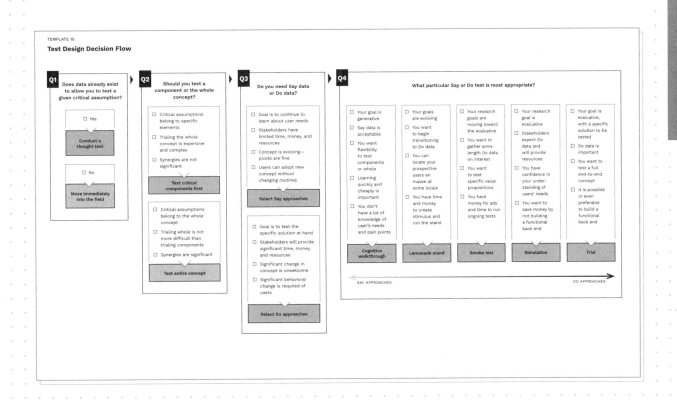

TEMPLATE 10

Test Design Decision Flow

Q1 Does data already exist to allow you to test a given critical assumption?

☐ Yes

[Conduct a thought test]

☐ No

[Move immediately into the field]

Q2 Should you test a component or the whole concept?

☐ Critical assumptions belong to specific elements

☐ Trialing the whole concept is expensive and complex

☐ Synergies are not significant

[Test critical components first]

☐ Critical assumptions belong to the whole concept

☐ Trialing whole is not more difficult than trialing components

☐ Synergies are significant

[Test entire concept]

Q3 Do you need Say data or Do data?

☐ Goal is to continue to learn about user needs

☐ Stakeholders have limited time, money, and resources

☐ Concept is evolving— pivots are fine

☐ Users can adopt new concept without changing routines

[Select Say approaches]

☐ Goal is to test the specific solution at hand

☐ Stakeholders will provide significant time, money, and resources

☐ Significant change in concept is unwelcome

☐ Significant behavioral change is required of users

[Select Do approaches]

Q4 What particular Say or Do test is most appropriate?

☐ Your goal is generative
☐ Say data is acceptable
☐ You want flexibility to test components or whole
☐ Learning quickly and cheaply is important
☐ You don't have a lot of knowledge of user's needs and pain points

[Cognitive walkthrough]

☐ Your goals are evolving
☐ You want to begin transitioning to Do data
☐ You can locate your prospective users en masse at some locale
☐ You have time and money to create stimulus and run the stand

[Lemonade stand]

☐ Your research goals are moving toward the evaluative
☐ You want to gather arms-length Do data on interest
☐ You want to test specific value propositions
☐ You have money for ads and time to run ongoing tests

[Smoke test]

☐ Your research goal is evaluative
☐ Stakeholders expect Do data and will provide resources
☐ You have confidence in your under-standing of users' needs
☐ You want to save money by not building a functional back end

[Simulation]

☐ Your goal is evaluative, with a specific solution to be tested
☐ Do data is important
☐ You want to test a full end-to-end concept
☐ It is possible or even preferable to build a functional back end

[Trial]

SAY APPROACHES ⟶ DO APPROACHES

4

Finally, we will summarize, in one page, all of the details of your test strategy, using the **Test Digest** [TEMPLATE 11, *page 94*].

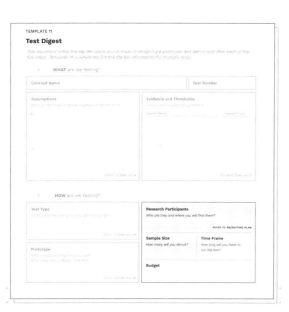

To do this, we first drill down to more detail on who your research participants will be and where you will find them. You have a target audience you want to learn from and test with, but you need a plan for locating and recruiting them. And if that plan doesn't work, you need a backup plan.

Create a **Recruiting Guide** to capture this information.

Creating a Recruiting Guide

What is it?	A short memo that notes your plans A, B, and C for recruiting your target audience. Where will you recruit? Will you intercept them at a conference? Will you send an email to try to get people to sign up? What is your strategy?
Why is it important?	Testing with the wrong audience can create faulty data and waste your time and resources.
How to construct it	Carefully note the characteristics of your target audience, how many people you want to talk with, and what channels you'll use to recruit them (e.g., stop them on the street, set up a stand at a conference, use a recruiting service, reach out on LinkedIn).
Expert tips	In many cases, it might be harder or take longer to identify and recruit your target audience than you expect. Having plans B and C allows you to immediately move into the next plan for recruiting, ensuring you can continue to run your experiment effectively and efficiently. Otherwise, you may be tempted instead to just talk to folks who are easier to recruit but aren't your target market. Having B and C already in place helps to avoid this—and you might have discovered that you require more time or money (to provide an incentive to thank users for their time, for instance). Talking about this as a team and with your stakeholders ahead of time is key.
Equity	Within your intended audience, pay attention to whether you are recruiting or using a diverse group of research participants. Is your research accessible to people without a computer during research hours? Is your research accessible to people of a variety of ages and socioeconomic levels? Is your research accessible to people with a variety of lived experiences related to the concept being tested? Who are you currently missing?

4

Next, we will fill in the remaining details in your **Test Digest** [TEMPLATE 11, *page 94*].

Log your progress in your **Progress Tracker** [TEMPLATE 1, *page 82*]:

☑ Choose between Say and Do tests

☑ Select a test type

☑ Design your test

Congratulations! You have just one more critical decision to make before launching your field work: the specifics of the prototype you will build. It's time to create of the right prototype for the test you have designed.

On to Step 4!

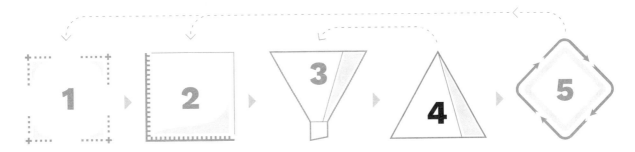

Building the Prototype

IN STEP 4 YOU WILL LEARN TO:

- Assess the level of fidelity needed

- Select prototype format

- Construct your prototype

RECALL THE ESSENTIAL components for experimentation that we highlighted in our introduction: hypothesis, intervention, and stimulus. Now that you have your concept framed as a testable hypothesis (Step 1) with prioritized assumptions and defined evidence (Step 2) and have selected your test type (Step 3), you are ready to complete the final component—your stimulus—by building the prototype that will elicit the data you need from your potential users.

You know the old adage "show, don't tell"? That's the essence of prototyping. It's the difference between your friend *telling* you about a new idea in words versus *showing* you a sketch of it before saying, "What do you think?" Prototypes can be paper-based, digital, or physical representations of concepts for services, products, processes, or experiences. We use prototypes as the provocation to confirm or disconfirm our assumptions. They help us gather accurate data about the desirability, feasibility, and viability of our concept.

Like test types, prototypes can take many forms. They usually start simple and get more complex as the desirability, feasibility, and viability of the concept being tested start to become clearer. In this step, we will explore the different formats prototypes can take, provide guidance on selecting the most appropriate prototype format for your situation, and offer practical tips to help you successfully construct your prototype.

As the Nike story illustrates, prototypes vary along many dimensions. These include the materials they are made of, the time and resources required for their creation, and the extent to which their appearance, functionality, and interactivity mirror a finished product or service. In this step, we will focus on the most commonly used **prototype formats.**

Seeing Is Believing

One of the most unusual—and compelling—uses of prototyping we have seen comes from Mexico, where MasAgro, a partnership between the Mexican government and agricultural groups, works with local farming communities to encourage the adoption of sustainable modern agricultural methods. MasAgro's creative use of prototyping demonstrates how experimentation can not only test—it can encourage buy-in and further implementation success. Because subsistence farmers' entire livelihoods rely on each year's harvest, they are understandably risk averse and loath to abandon traditional tried-and-true methods for new ones, even those aimed at raising their income. MasAgro doesn't just tell farmers about the benefits of change—it *shows* them. The government works with respected community leaders to create trial plots with rows of crops—some using modern methods and others traditional—planted side by side so farmers can see for themselves, at the end of the growing season, the difference the new methods make. As a result, MasAgro has succeeded where other organizations with similar aims have failed: more than 40 percent of participating farmers have adopted at least one innovation—an extraordinary success rate.

Prototype Formats

Storyboards are sequences of visuals showing how users interact with the product or service. A storyboard can be printed on paper or drawn in an electronic format.

Posters are two-dimensional descriptions of the key elements of a product or service.

Pitch decks are presentations that show the intended product or service and its value proposition. They are similar to a storyboard in that they walk users through a story, but usually in a slide deck rather than a series of illustrations.

Mock-ups are physical or digital representations of elements of the product or service that allow users to interact with the offering. These can be created at different levels of fidelity.

Minimum viable products (MVPs) have real-looking front ends with basic back-end functionality. They mirror the look, basic operation, and interactivity at a level closer to the finished product.

Assessing the Needed Level of Fidelity

Many of the decisions you need to make as you select your prototype format are related to *fidelity*, a central concept in prototyping. Prototypes range along a continuum from low to high fidelity. High-fidelity prototypes look and act very real; low-fidelity prototypes do not. Experimenters are often tempted to build the most complete prototype they can. Instead, **think about starting with the *simplest* prototype that will let you accomplish the job at hand**—testing your prioritized assumptions. Counter-intuitively, simple prototypes have been demonstrated to elicit more accurate feedback than polished ones.

Core aspects of fidelity relate to a prototype's form, function, and interactivity. Let's look at each in turn:

- Form refers to how real a prototype *looks*. This can range from simple sketches (low fidelity) to physical or digital versions of products or services that appear close to what the finished product will look like (high fidelity).

- Function refers to how a prototype *works*. Does it actually perform the function that the final product or service is meant to do? For instance, you could create a very realistic-looking vacuum cleaner that still does not pick up dirt, giving it high form but low functionality. Some prototypes, such as storyboards, make no attempt to "work"—they are just on paper. Others, like MVPs, possess much of the critical functionality of finished products.

Evolution of the *Easykicks* Prototypes

Peer Insight utilized a variety of prototypes during the *Easykicks* experimentation cycle. Initially, their goal was to figure out whether people would even want a shoe subscription service. If they did, *why* would they want it? Because the concept was so early in its experimentation journey, they didn't need a sophisticated, operational prototype to test basic assumptions about desirability. They sketched the idea in a simple, paper-based prototype, creating just a ten-frame storyboard (shown in Step 1) to help answer basic questions. Later, as they iterated their concept, they moved to in-depth testing of key components, such as the willingness to return old shoes described in Step 3. The return of shoes was a critical part of Nike's interest in the idea—evidence of their commitment to sustainability—and they had already invested in capital equipment to do the reprocessing. Peer Insight needed to address all three assumption conditions around this component of the concept: whether people wanted to return their old shoes (desirability), whether they actually would return them (feasibility), and whether the expense of processing returns would not be too high (viability).

To assess these three assumptions, Peer Insight created a physical mock-up as part of a simulation, giving people a bag at shoe purchase so they could mail their old shoes back. Later, they used digital prototypes in smoke tests to test the specifics of the value proposition and then a simulation and ultimately a trial to test the end-to-end experience.

The Fidelity Continuum

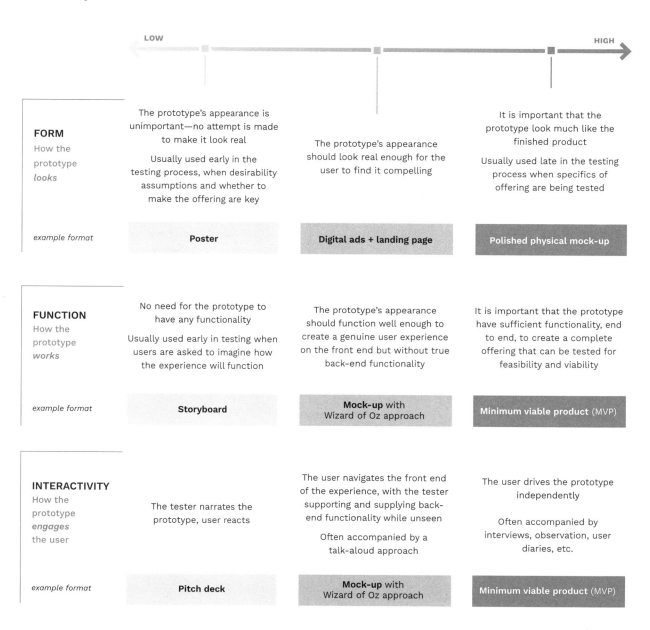

LOW   HIGH

FORM
How the prototype *looks*

The prototype's appearance is unimportant—no attempt is made to make it look real

Usually used early in the testing process, when desirability assumptions and whether to make the offering are key

The prototype's appearance should look real enough for the user to find it compelling

It is important that the prototype look much like the finished product

Usually used late in the testing process when specifics of offering are being tested

example format

Poster

Digital ads + landing page

Polished physical mock-up

FUNCTION
How the prototype *works*

No need for the prototype to have any functionality

Usually used early in testing when users are asked to imagine how the experience will function

The prototype's appearance should function well enough to create a genuine user experience on the front end but without true back-end functionality

It is important that the prototype have sufficient functionality, end to end, to create a complete offering that can be tested for feasibility and viability

example format

Storyboard

Mock-up with Wizard of Oz approach

Minimum viable product (MVP)

INTERACTIVITY
How the prototype *engages* the user

The tester narrates the prototype, user reacts

The user navigates the front end of the experience, with the tester supporting and supplying back-end functionality while unseen

Often accompanied by a talk-aloud approach

The user drives the prototype independently

Often accompanied by interviews, observation, user diaries, etc.

example format

Pitch deck

Mock-up with Wizard of Oz approach

Minimum viable product (MVP)

- Interactivity looks at the way users *engage* with the prototype. How authentic a user experience does it provide? At the low end, a poster with a photo or sketch and description of a new vacuum cleaner makes no attempt to give users an authentic experience. But even a high-form, high-function version of our vacuum cleaner is not highly interactive if it is handled only by those conducting the test and a user never actually experiences what it feels like to vacuum with it.

Think of a **continuum** from low to high that captures these three elements. Here are some examples of where different prototype formats might fall.

Not surprisingly, the time and resources required to create a prototype will tend to increase as its fidelity in form, function, and interactivity increase.

Type of test also matters—different test designs have differing requirements for form, fidelity, and interactivity. A cognitive walkthrough, done early in the testing process, usually requires only a simple storyboard with low form, function, and interactivity, like the one Nike used at the start of testing *Easykicks*. On the other hand, a test like SWR's *Concierge* trial utilized an MVP prototype, whose functionality, form, and interactivity were high.

Selecting a Prototype Format

With all these factors at play, choosing the right prototype format for any given test may seem daunting. Keep in mind that multiple formats will be used over the life of an experimentation cycle. At first, we want to select the fastest, cheapest prototype that will do the job it needs to do. We rarely see teams underinvest in prototypes that are not sophisticated enough. Usually we see the opposite—overinvesting in prototypes that are far higher in fidelity and cost much more in time and resources than is needed at that point in the concept's life. At its simplest level, the right prototype is the one that allows you to test your assumption and to gather the data you need to confirm or disconfirm it.

Let's look at the most common prototypes that are paired with different types of tests.

Prototypes Suggested for Cognitive Walkthroughs or Lemonade Stands

When doing a cognitive walkthrough or a lemonade stand, you're likely in the early stages of experimentation. You often have little money or don't feel comfortable spending a lot. Your assumptions are mostly around desirability: Do people have the problem you think they have? Are they seeking the benefit the concept solves for? Are they remotely interested in the proposed solution?

Common prototypes used at this stage are storyboards, pitch decks, and posters. These two-dimensional prototypes require little time and money to prepare, are good for testing basic desirability assumptions, and can be administered in person (on paper) or virtually (digital). If you are prototyping an experience instead of a product, storyboards have an advantage. If you want to ask users to quickly compare different concepts, posters work well. If your audience is a business (rather than an individual consumer), pitch decks are often most valuable.

Prototypes Suggested for Smoke Tests or Simulations

If your test type is a smoke test or simulation, you likely have some proof of a need for your concept, and so are willing to spend more money. Your assumptions around desirability, feasibility, and/or viability are more specific than earlier in the experimentation process. Testing them requires more functionality and interactivity—and often a more polished form. What does your solution need to look like exactly? Can you execute the solution in the way that you envision? Can you do it at a low enough cost that someone might be willing to pay you enough for it? If the answers to these questions are yes, you are ready for a smoke test or simulation.

Common prototypes with smoke tests or simulations are digital and physical mock-ups, and digital ads with landing pages. Though they are more costly and time intensive to build than a two-dimensional prototype like a storyboard or poster, they allow you to test specific value propositions rather than general

Prototypes for Lemonade Stands and Smoke Tests at PMI

The Peer Insight team, working with PMI to assess which of the four concepts should be identified to moved forward, designed and executed two rounds of tests, using multiple prototypes. As we discussed in Step 3, they operated lemonade stands at a PMI global event, displaying paper-based poster prototypes for each of the concepts. They also conducted smoke tests, using digital prototypes consisting of LinkedIn ads connected to landing pages. From this research, they decided to move two concepts forward—*Career Navigator* and *Snippets*—and created more sophisticated simulations for the next round of testing. The *Hive* and *Spot* concepts were shelved.

desirability. They offer higher functionality to test feasibility assumptions and higher interactivity that allows you to refine the user experience and acquire more robust Do data. Costs and time invested can be minimized by using a Wizard of Oz approach that manually mimics back-end functionality. Digital ads and landing pages are particularly good at assessing value propositions. Digital mock-ups using wireframes can make software solutions feel real without expensive coding, just as physical mock-ups (like sheets hung to mimic walls at a medical clinic to test examining room sizes for functionality) can make physical environments feel real.

Prototypes Suggested for Later-Stage Simulations and Trials

If your test design includes later-stage simulations or trials, you are likely further along in your experimentation journey or you need to move fast. You likely have good evidence that users want your concept and good directional data on the two or three main features it should contain. You also have some data indicating long-term viability. Having such data gives you confidence to spend additional time and resources to move into the actual building process. Do all the features and components work together seamlessly? What is the level of effort to fulfill the delivery of this product/service? Does it show signals it can scale? How will users consume the product/service over time? Why?

Commonly used prototypes here approach high functionality, either created via a Wizard of Oz approach or actually built into an MVP.

Prototypes for Simulations at PMI

Having narrowed PMI's concept to *Snippets* and *Career Navigator*, the Peer Insight team turned to the question of learning more about the features and business model of each. They created simulations for each that used digital mock-ups as prototypes, employing a Wizard of Oz approach. Each prototype had a front-end user interface that was intended to feel real but lacked an operational back end. After testing these digital mock-ups with users, Peer Insight felt confident that they understood what to make in an MVP, supported by their robust and triangulated research plan.

Evolution of the *Concierge* MVP at SWR

DK&A's decision to move quickly to a trial of the *Concierge* concept at Basingstoke Station required an MVP that functioned on both the front end and the back end. The MVP prototype they created had two key components: (1) a pop-up that acted as the *Concierge* desk to provide the user interface and (2) employees trained in the new concept to provide the functional service. For the pop-up, they found readily available off-the-shelf materials that could be quickly imprinted with Basingstoke Station specifics. The total cost of materials needed was about £400. Existing interactive digital display boards were repurposed for inclusion as part of the prototype. To train employees to deliver the service, DK&A worked with SWR's Learning and Development team to design and deliver a training program, complete with supporting videos. As they moved to the second round of testing and faced the question of how to scale the *Concierge* concept throughout their network of stations, the MVP itself had to evolve. Feedback on the initial MVP was positive but suggested that some passengers had seen it as a marketing or sales desk, rather than a help desk, so had avoided it because "it looked like someone was trying to sell something," as one passenger explained. The redesigned MVP fitted into a larger array of station types, both large and small, and had signage that made clear that it was a help point. The educational component was also expanded, with Basingstoke employees (already experienced from the first test) acting as coaches and trainers.

ASSIGNMENT

Assignment

Now it's your turn again!

1

Use the **Prototype Format Selection Tool** [TEMPLATE 12, *page 96*] to guide your selection of a prototype format.

2

Once the prototype format for your test has been identified and aligned with the job it needs to do, it is time to **make the actual prototype** that users will interact with.

Below, we offer some details on materials and recommended steps to follow from those who have real-world experience designing hundreds of prototypes for a variety of teams.

To-do list for building your prototype:

- ☐ Begin by reviewing Steps 1–3
- ☐ Sketch an initial design of the prototype
- ☐ Begin an outline of your research guide (more on this shortly)

- ☐ Select the software or physical materials you need and have access to
- ☐ Create the drawings for your storyboard, if needed
- ☐ Consider the use of digital whiteboards like Mural, Miro, or Jamboard if you plan to present your prototype virtually. These will be easy to view and allow for real-time feedback from users
- ☐ Create simulations with physical materials or electronically in FullStory, Figma, InDesign, Hotjar, or other applications
- ☐ Walk through your prototype with your team, and continue to iterate as the prototype comes to life

Expert Tips

Consider Your Audience
Put yourself in the shoes of one of your research participants: make the prototype simple and accessible to your audience (e.g., easy-to-follow instructions, no small font, ADA [Americans with Disabilities Act] compliant, legible printed materials).

Consider Any Privacy Concerns
Are you making any data requests that need permission granted? Do you need your research participants to sign any waivers, either legal ones or those required by any stakeholders you're working for?

Consider Accessibility
Does your software need a sign-in? Does your prototype need to be accessible on a mobile device—Android and Apple—or will your users be expected to be at a desktop computer?

ASSIGNMENT

3

You have your prototype, and you're ready for action—right? Not so fast. Before you move to Step 5, one more task awaits: creating guidance for you and your team to ensure that they use the prototype well—gathering the right data from the right people and ensuring that each researcher does this in a consistent way. This guidance comes in the form of two documents: the recruiting guide that you created during Step 3 and a **Research Guide** that you will create now. Your recruiting guide helped you select your users for the test. Your research guide is your manual for using the prototype with those users; it helps ensure that you collect the data you need.

These guides work in tandem and inform each other. They also ensure that everyone administering the test is on the same page, and that all testers have been briefed on the "why" and "what" of the test and prototype design. This background knowledge allows them to pivot on the fly as necessary.

To create your research guide, start by referring back to Step 2 and your assumptions and evidence to review what data you need to gather through your prototype. Use those as reference points as you put together your research guide. The adjacent list shows the most common elements used in a research guide. You may not need all of them.

Creating a Research Guide

Role of researcher and role of research participants

Outline the specific activities or protocol any researchers (teammates) will need to perform during the test. For example, with a cognitive walkthrough, you'll outline how you will interact with both the prototype itself and the research participant. For smoke tests, you'll outline how often to check on the ad performance and when to iterate.

For the research participants, or users, you'll note how you'll interact with them through the prototype—do they need an email from you with instructions? Any prompts or interactions with the research team? Outline these steps ahead of time to be prepared.

Interview guide and/or text

You want to make the most of your time with users, so a list of questions for the interview will help you stay organized and focused. If you're running a smoke test, you'll need to carefully draft your copy using your assumptions to launch your landing page and ads. For the simulation, what are all the actions your research participant needs to take or behaviors you need to observe to get the data you need?

With most prototypes and tests, there are always opportunities to probe further if a user responds in a particular way that you want to learn more about. So, ahead of time, identify areas for potential probing.

Privacy notification and/or consent forms

Always ensure you are following relevant privacy and data regulations. You may need to note how you're using the data, where it is being stored, and whether it will be anonymous. You might also need to have research participants sign a consent form.

Why take the time to create this?

A research guide helps you be consistent across testers and ensures that you answer the questions you want to answer, and not ones you don't.

Creating a research guide often highlights tweaks you need to make to the prototype— maybe an extra feature in a simulation, or an extra frame in a storyboard.

It's easy and tempting to skip this step—you're excited to get out and test. But do it anyway: even if you don't use it verbatim, you'll be more focused going into the test.

ASSIGNMENT

Research Guide Prepared for SWR Employees

....

Seeing Through the Eyes of the Customer

Observe the station environment
Try to look at the station with fresh eyes. If a friend or relative came here, what would they think or feel? How are people moving around the station?

Observe the customers
Watch the customers' behaviors and their expressions. What is their mood?

What are we hearing from customers?
What are people asking staff about at the station? What are they confused or concerned about? Are there things other than train times that they want?

Actively approach and chat to customers
Ask passengers: Why are they here today? Why did they take the train rather than traveling a different way? What is their experience of the station? What is good, bad, surprising? What would encourage them to use the train and station more?

....

Collect a Record of What We See and Hear

• Take our phones with us so we have them on hand, and a paper and pen too if desired
• Share notes and photos via the WhatsApp group

Remember to ask permission before taking photos of people.

With your test plan, prototype, and guides in hand, you are ready to rock and roll! Log your progress in your **Progress Tracker** [TEMPLATE 1, *page 82*]:

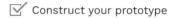

☑ Assess the level of fidelity needed

☑ Select prototype format

☑ Construct your prototype

Now the moment we have been waiting for . . .

On to Step 5!

ASSIGNMENT

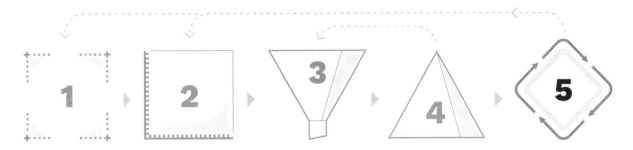

Executing. Analyzing. Iterating.

IN STEP 5 YOU WILL LEARN TO:

- Audit and pretest your design

- Run your test

- Share your findings

- Iterate your concept

NOW THAT YOU have selected a test, built your proto-type, and created recruiting and research guides, you are ready to move from planning the *how* of your experimental process to actually learning from the field. Let the experiments begin!

As with almost every stage we have talked about thus far, *iteration* is a key component of Step 5. Not only do we expect to iterate through different types of tests and fidelities of prototypes over the life cycle of testing a concept, we will also iterate *within* a given test—auditing the test design, pretesting it, running the test, analyzing the findings, sharing them, iterating the concept, and designing the next test. So, it is not just that the *concept* itself gets refined—the *process of testing* will also evolve. Often that adjustment will need to occur in real time, on the fly—so preparation and contingency planning are especially critical.

In this step, we will focus on the design and execution of a single test, but keep in mind the larger context that includes the multiple tests in your overall plan for learning through the experimentation process.

Let's look in more detail at Step 5's **iterative process** and the activities it contains.

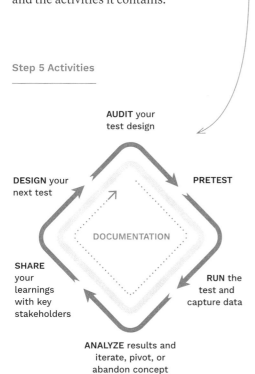

Step 5 Activities

AUDIT your test design

DESIGN your next test

PRETEST

DOCUMENTATION

SHARE your learnings with key stakeholders

RUN the test and capture data

ANALYZE results and iterate, pivot, or abandon concept

The Importance of Documentation

One important activity throughout Step 5 is documenting your process. You may be tempted to put off thinking about documentation until after you have run your test and collected your data. Don't! Documentation should be happening in parallel with the other activities in Step 5. The credibility of your findings rests on your ability to demonstrate the soundness of your process. If your findings run counter to what key stakeholders expect or want to see, they may attempt to discredit your testing design and execution rather than accept your findings. So, think in advance about how you will share your process and your findings with them. Tag

Begin with the End in Mind

Stephen Covey famously advised us to begin with the end in mind. Nowhere is this truer than in documenting the integrity and rigor of the process. You may find it useful to use an old consultant's trick: write the future slide deck you envision presenting to your key stakeholders after the study is complete, if all goes as you hope it will. Think of it as the companion to the pitch deck you presented to sell the concept—but with evidence to support it, rather than just your assumptions. Of course, for now the actual evidence numbers will be blank (you haven't gathered your data yet), but you can think about the headlines, the data, and the flow of the story that people will find convincing and compelling. Be equally prepared to abandon your plans when the surprises start—but as General Dwight Eisenhower said about the importance of planning when the fighting starts: "Plans are worthless, but planning is everything."

compelling quotes and stories as you go. Make note of what surprises you and gets disconfirmed, not just what confirms your beliefs. The integrity and rigor of your testing process and the narrative arc of your findings must capture your learning in ways that make your findings clear and compelling to your audience.

The templates you created in Steps 1–4, recording your decisions as you made them, should make it easier to establish and explain your choices throughout the process.

Auditing and Pretesting Your Design

Having laid out your documentation strategy, take one final pause before jumping into the field, to *audit* and *pretest* your design to increase the likelihood of producing credible results. When you are ready, you can begin your assignment for Step 5 by working through the **Test Audit Checklist** (TEMPLATE 13). Once you are confident in your design, you are ready to take it out for a test drive—but not yet with actual users. Begin gently testing its functionality with a group of sympathetic stand-ins for your real users. For instance, DK&A's pretesting consisted of doing cognitive walkthroughs with the three SWR station managers, acting as proxies for customers and employees. You are looking for people who are committed to helping you improve your test. Ask them to walk themselves through the prototype, asking you questions (rather than the other way around). Do they understand the essentials of your value proposition? Is your data capture plan working? Are you collecting the information you need to confirm

or refute your assumptions? Pretesting focuses on testing the functionality of the test *itself* more than gaining feedback on the assumptions the prototype is testing. Pretesting with friends, family, or proxies helps ensure that your findings will not be biased by how your test is conducted.

Once you have completed this test drive, you are ready for the open road.

Running the Test and Capturing Your Data

Having a good data capture strategy is critical, particularly when your methods include real-time interactions like cognitive walkthroughs and lemonade stands. There are many types of apps and technology to aid in capturing and analyzing data. They typically fall into two categories: tools for capturing raw data (for example, Otter.ai, notetaker, WhatsApp, Hotjar) and tools for sorting/synthesizing the data (for example, Google Analytics, Airtable). Find technologies that integrate via APIs (mechanisms that enable different pieces of software to "talk" to each other) to minimize any manual connections and having to cut and paste data from one app to another.

For the PMI project, Peer Insight used a variety of tools. In addition to tried-and-true workhorses like Google Analytics (to capture web data to note user behavior) and Excel (to store and sort data once it's captured), they found Data Dashboard (a simple table that captures high-level, simplified data) helpful for providing an "at a glance" update on key product metrics—including number of users, behavioral data, and category data—for stakeholders and executives. The team reviewed it daily to share trends, individual user reactions, and content. They also used Notetaker (a voice-enabled note-taking app) integrated with Airtable (an easily manipulated aggregation/analysis tool) to quickly capture and answer questions about users, product content, and features and to tag quotes for later use.

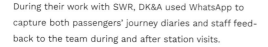

Using WhatsApp at SWR

During their work with SWR, DK&A used WhatsApp to capture both passengers' journey diaries and staff feedback to the team during and after station visits.

For passengers, they set up a WhatsApp group as a follow-on to interviews, with the individual and the DK&A team as members. They asked people to report on their journeys from booking to completion. The basic instruction was to record their experiences as they happened—taking photos or videos and jotting down notes about their thought processes and experiences. The team was able to ping them with questions if anything struck them as interesting, surprising, or contradictory.

WhatsApp was also an easy-to-use, familiar app for frontline staff. Because they were busy and often on the move, they didn't necessarily have easy access to laptops. WhatsApp was free, everyone had a smartphone, and most had used WhatsApp or could learn it. So, DK&A set up a group for each station in order to keep in touch with staff the team met during their station visits.

New supportive software applications are constantly being introduced, and even mainstream social media apps like Instagram and WhatsApp can be useful, as DK&A demonstrated in their work with SWR.

Analyzing Results and Iterating Your Concept

The moment has come—you have results! Now what do you do with them? How do you assess what they mean? It all goes back to those critical assumptions that you took the time to lay out in Step 2, built into your test plan in Step 3, and fashioned into your prototype in Step 4. Use these as your guide.

Let's return to Whiteriver and check how they used what they learned from their initial testing. The Whiteriver Team designed and ran a four-day simulation, arranging for an experienced physician to greet each emergency room arrival to quickly triage patients as urgent or nonurgent. They focused on a single point of evidence: how many visitors left without seeking treatment. The results were impressive. The percentage of arrivals abandoning Whiteriver's emergency room without treatment was reduced from 18 percent on the control days to less than 2 percent during the test. The team then took this difference and calculated a rough estimate of the effects such a reduction would have on the hospital's finances, starting with existing performance indicators (in this case, wait times, costs, and lost revenues). They knew that the hospital was losing money when potential patients left without being seen, but they had to quantify the effect to demonstrate the value of their ideas to administrators. Because Whiteriver was so far from major medical centers, most seriously ill patients were flown to Tucson. But since so many emergency room visitors with minor problems left and then returned more seriously ill, catching patients in the middle zone, when they could be admitted, would lead to higher bed-use efficiency with more income from the federal government. Beginning with the unrecovered revenue from the 20 percent of potential patients who left without being seen, the team used the hospital's average daily income per patient to calculate slightly over $3 million in uncaptured revenue. In addition, basing their calculations on the average patient, they estimated that the 8,000 patients who left would also have needed some $2 million in pharmaceuticals. Finally, the mean cost to transfer patients from Whiteriver to bigger hospitals was $1,700 per patient. Although 1,000 patients were flown elsewhere in an average year, the team decided to use a conservative estimate that roughly a third would still require transfer, which put that cost savings just over $1 million. In total, the team came up with $6 million savings against a cost of $150,000 to redesign the emergency department to separate patients dealing with basic medical concerns from those actually needing emergency services.

When the team presented the figures, hospital leadership found the fast-track idea compelling. Even allowing for a wide margin of error in the team's estimates, both the benefit to patients and the return on investment were clear.

SWR ran two rounds of trials to test the *Concierge* concept. During the first trial, the DK&A team went to the Basingstoke Station to observe and informally interview customers and staff to assess whether the assumptions concerning passenger desirability and staff enthusiasm were supported, and to gauge how well the concept worked using existing staff. They recorded over six hundred interactions during this first trial and conducted intercept interviews with passengers and staff. The team concluded that their make-or-break assumptions about the desirability of the new concept were supported, having witnessed enhanced staff–customer interaction, improved satisfaction and increased self-service, enthusiasm from staff, and interest from other station managers and

The Leisure Travel Component of the *Concierge* Concept

As originally conceived, the *Concierge* had three discrete components that DK&A tested: a giveaway COVID test kit, the concierge service itself, and a leisure travel promotion. The leisure travel promotion idea offered an interactive advertising board with staff recommendations for local leisure trips, aimed at stimulating conversations and promoting leisure journeys. During the initial test at Basingstoke, despite prominent placement of the interactive ad boards and attractive content, awareness levels were almost zero. The test revealed that leisure promotion was badly timed, as commuters were concerned about their immediate travel, not future plans. The leisure travel component was eliminated from the concept moving forward.

colleagues. Records of self-ticketed revenues (sales from ticket vending machines, or TVMs) also increased 7 percent during the trial. However, not all components succeeded, and the leisure travel component did not meet desirability thresholds. Feedback on the *Concierge* name was also mixed, so the concept was renamed the *Welcome Host* for the next trial.

Having established desirability, DK&A embarked on a more rigorous testing approach in the next trial, aimed at scaling the renamed *Welcome Host* concept across a more diverse set of stations and customers. The new prototype contained enhanced technology for operators and greater ease of setup and training. The second trial lasted six weeks and was conducted in

Welcome Host Research Assumptions and Results

	Assumption	Test 1	Test 2	Results
DESIRABILITY	Customers find the service helpful	Y	Y	Over 13,000 interactions were recorded and very positive customer feedback about the service was received.
	Customers' basic questions easily answered	Y	Y	Common customer queries were quickly answered, including requests for information about journeys, tickets, and directions.
	Reduced queues for customers	Y	Y	*Welcome Hosts* approached customers when queues built to answer questions or help them use the Ticket Vending Machines effectively. As customer numbers rise, colleagues feel that the impact of the *Welcome Host* will grow.
FLEXIBILITY	Adapts to the needs of different stations	N/A	Y	Stand branding, size, and location were varied to fit station layouts and needs. The service particularly worked in busier stations with leisure users. Staff in smaller stations felt the experiment was too soon.
	Uses existing skills without costly training	N/A	Y	Effective in-house Customer Experience "refresher" training attracted forty-three SWR volunteers from multiple roles with additional informal peer-to-peer training in some stations as the project progressed.
	Can be run with existing staffing levels	Y	Y	More than 60 current staff were involved in the Test 2 experiment, with many stations setting up a *Welcome Host* roster. Most employees enjoyed the work and would like to continue the role full- or part-time.
	Uses available equipment (such as digital totems [freestanding digital information displays])	Y	Y	SWR totems and tablets were made available to provide useful support information about journeys and the local area. Totems in particular were useful for the Hosts and for some customers to self-serve.
VIABILITY	Technology integrated to increase self-service	Y	Y	In the Test 1 experiment the proportion of Ticket Vending Machine to Ticketing Office sales increased by 7 percent. There was no measurable increase in the Test 2 experiment, but *Welcome Host* staff felt the service benefits would increase as passenger numbers grow.
	Potential to free up space for increase retail	Y	Y	Part or all of the Ticketing Office space could be used for commercial/retail opportunities if *Welcome Host* has a permanent stand out in the hall with secure storage, access to support information, and the ability to sell and read tickets.
	Potential to self-fund	Y	N	Self-funding would require the cost of permanent stands in the ticket hall plus hall redesign to be offset by extra income opportunities (from increase in efficiency, Test 1 volume of sales, and extra retail space).

eight stations. Data collection included surveys, data logs, and qualitative user interviews. Transactions were recorded manually by *Welcome Host* employees. Again, results demonstrated that the warm welcome and quick support were highly valued by customers. On the staff side, 80 percent of employees involved in the *Welcome Host* concept reported enjoying the trial and felt that it improved the customer experience, and 62 percent wanted to retain their new role rather than return to their previous one. SWR captured the **evidence from the two rounds of trials** against the assumptions they had identified for desirability, feasibility, and viability.

Based on this evidence, a series of recommendations were made to SWR management following the second trial:

1. Continue the existing trials at stations already using the *Welcome Host* to capitalize on momentum and increase the data gathered.
2. Expand the training model to include peer-to-peer training in addition to formal skills training.
3. Revise ticket hall layouts to focus on customer flow, experience, and retail/community space.
4. Run a future pilot with the *Welcome Host* replacing the ticket office completely.

Now let's turn to PMI and look at how their learning progressed as Peer Insight worked through successive rounds of tests and prototypes. Round 1 testing resulted in the elimination of two concepts, *Hive* and *Spot*. Peer Insight determined that *Hive* failed to meet threshold levels of desirability. Those who did indicate interest indicated split desires: some wanted mentoring and others desired quick Q&As. On the other hand, respondents loved the *Spot* concept, and it easily passed desirability thresholds. But when the team explored feasibility assumptions, they observed a significant discrepancy between the benefits that users wanted and the time they were willing to commit. So, these two concepts were abandoned and only the other two—*Snippets* and *Career Navigator*—moved on to the next round of testing.

For *Career Navigator*, a key metric since day 1 had been the percentage of users who persisted through the entire assessment. Peer Insight set a threshold of 40 percent. When they tested this using a simulation, the resulting 70 percent of users persisting far exceeded this threshold. But they had an aspirational goal of 100 percent and wanted to understand why the 30 percent who dropped out did so. Through interviews and journals in Round 1 testing, they identified that many of these users started the assessment but did not feel prepared at that time to provide all the necessary information requested on previous career experiences. So, in refining the concept, Peer Insight extended the *Career Navigator* experience to include a series of nudge emails, to remind these users to return and complete their assessment.

As new questions emerged, the Peer Insight team designed specific tests for particular components in depth. During the trial of *Career Navigator*, the team observed that users in the premium (paid) version valued the action plan feature highly. Interested in gathering more information on the value and content of the action plan specifically, they ran a one-month trial where *all* users got the premium features. They did this to maximize their ability to address assumptions related to this key feature.

In some cases, subsequent testing disconfirmed findings from earlier tests. During the early cognitive walkthroughs of *Snippets*, for instance, video was a highly requested content format (versus audio or text), but in later simulations, the quantitative data showed low video consumption. As the team explored this surprising finding, they noticed that, in the qualitative data received from users' audio voiceover as they worked through the simulation, users noted that they weren't always in a position to watch a video, though they still preferred that format in theory.

Sharing Your Findings with Key Stakeholders

When the time comes to share what you've learned with your key stakeholders, your investment in documenting the testing process at every step pays off. In sharing both your journey and your findings, pay particular attention to the power of quotes and stories from your database to bring your findings to life and emphasize key points. Use language and examples that are meaningful and will resonate with your audience. *They* are now the user that you are designing with and for—think about their needs and what will create value for them.

The Peer Insight team created a **set of graphics** to share with PMI leadership to capture the flow of the learning plan they executed.

In another example, DK&A shared details on specific findings, such as **employee reactions**, with senior management at SWR, using quotes and images to make them compelling.

Colleagues from V1 and V2 Trials Would Want *Welcome Host* to Continue

TRIAL CONTINUATION

"There were seven staff involved (in V1 trial). Three of them have asked on a weekly basis when they can carry on the trial."

Station Manager, Basingstoke

CUSTOMER DISAPPOINTMENT POST-TRIAL

"Some customers were disappointed when it (the *Welcome Host*) was gone, especially just after it stopped."

Customer Operator,

Basingstoke V1 trial

TRIAL CONTINUATION

"It would be great to have this for when the Wimbledon tennis is on. It's expected people will be allowed to come to watch, but only half the usual numbers. That's still a lot of people who will be traveling."

Customer Service Assistant, Wimbledon

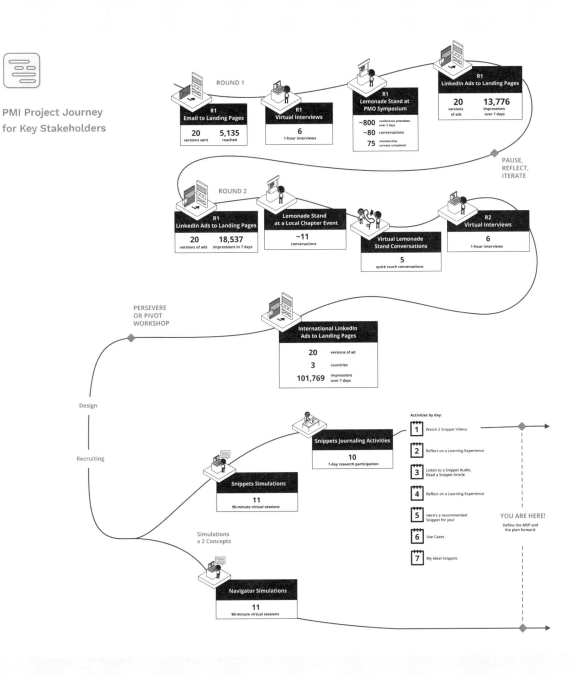

PMI Project Journey
for Key Stakeholders

ROUND 1

R1
Email to Landing Pages
20 — versions sent
5,135 — reached

R1
Virtual Interviews
6
1-hour interviews

R1
Lemonade Stand at
PMO Symposium
~800 — conference attendees over 2 days
~80 — conversations
75 — membership surveys completed

R1
LinkedIn Ads to Landing Pages
20 — versions of ads
13,776 — impressions over 7 days

PAUSE,
REFLECT,
ITERATE

ROUND 2

R1
LinkedIn Ads to Landing Pages
20 — versions of ads
18,537 — impressions in 7 days

Lemonade Stand
at a Local Chapter Event
~11 — conversations

Virtual Lemonade
Stand Conversations
5
quick touch conversations

R2
Virtual Interviews
6
1-hour interviews

PERSEVERE
OR PIVOT
WORKSHOP

International LinkedIn
Ads to Landing Pages
20 — versions of ad
3 — countries
101,769 — impressions over 7 days

Design

Recruiting

Snippets Journaling Activities
10
7-day research participation

Snippets Simulations
11
90-minute virtual sessions

Simulations
x 2 Concepts

Navigator Simulations
11
90-minute virtual sessions

Activities by day:

1 Watch 2 Snippet Videos

2 Reflect on a Learning Experience

3 Listen to a Snippet Audio,
   Read a Snippet Article

4 Reflect on a Learning Experience

5 Here's a recommended
   Snippet for you!

6 Use Cases

7 My Ideal Snippets

YOU ARE HERE!
Define the MVP and
the plan forward.

Designing Your Next Test, or *When are we done?*

Sadly, the question of how much experimentation is "enough" does not have an easy or obvious answer. It's not only about how much data you have collected. Budgets, timelines, and stakeholder expectations all factor into the decision about when to move from experimentation into commercialization. Are there still important unanswered questions about key assumptions? New ones will continue to emerge as you proceed through your learning journey. The process may never feel complete, yet you will need to move on at some point, probably before you feel ready. As architect Frank Gehry once explained: "You are never done. But eventually you run out of time and budget and so you move on."

It's easy to get stuck in the experimentation process and let a search for complete information get in the way of progress. Try to avoid the temptation to keep running a test in the hope of finally getting the data you want; at the same time, ensure that you've run it long enough to capture variation and establish clear communications with your target users. We suggest you frame your decision around stopping points by considering if you have *enough* directional data versus having *complete* data (an impossible feat).

Think in terms of *confidence levels* as you assess the question of how much more testing you need or want to do. Tests (especially early ones) are not about proof; they are about increased levels of confidence in your concept. As we increase our investment in a concept, we want to increase our confidence in it. We are in the game of *removing doubt*. As philosopher William James noted, the opposite of believing something to be true is not disbelieving it—it is doubting it. Vanquishing all doubt is impossible in an uncertain world. Our aim is to reduce it in proportion to our increased investment. Entrepreneur Alberto Savoia suggests thinking in terms of likelihood categories. He asks, "How confident am I that my assumption is true?" and uses five categories of response: very likely (90 percent confident), likely (70 percent), 50/50 (50 percent), unlikely (30 percent), or very unlikely (10 percent).

Triangulation of data sources and methods is the surest route to improving confidence that you are directionally correct. PMI gathered both qualitative and quantitative data using cognitive walkthroughs, lemonade stands, smoke tests, and simulations. SWR relied on customer observations and interviews, mystery shoppers, interaction counts, sales data, and feedback from station managers and employees using archival data (from voice-of-the-customer surveys) during cognitive walkthroughs and trials.

Once you and your key stakeholders agree that you are "confident enough" to move to the next level of investment and experimentation, you are ready to complete this cycle in the experimentation process and continue your larger learning journey by designing your next test. As you run multiple tests, it is important to document the changes to your concept. We recommend using a template like the one provided in the **Concept Test History** (TEMPLATE 14, *page 87*).

Assignment

And so, we reach your final assignment—running your test and capturing the results!

1

Be sure to have a process in place for documentation and think ahead to how and what you will share with your key stakeholders after the test is run. Then run through the items in the **Test Audit Checklist** [TEMPLATE 13, *page 97*].

2

When you are ready to go, run your test and capture the changes to your concept in your **Concept Test History** [TEMPLATE 14, *page 98*].

TEMPLATE 13

Test Audit Checklist

- ☐ The Concept Snapshot for this idea specifies who you have designed it for, how it meets their needs, and why it is differentiated from other options already available.
  If not, return to STEP 1 and refine your snapshot entries.

- ☐ The critical assumptions to be tested in this round are clear.
  If not, return to STEP 2 and review the assumptions to be tested.

- ☐ The evidence you will gather aligns with the assumptions to be tested.
  If not, return to STEP 2 and review the evidence needed and targets assigned.

- ☐ You have established clear threshold and aspirational targets.
  If not, return to STEP 2 and establish these targets.

- ☐ You have selected the test type best suited to your situation. If not, return to STEP 3 and retrace your progress through the questions posed by the decision flow.

- ☐ You have recruited the right people to test with.
  If not, return to STEP 3 and review your recruiting guide and instructions.

- ☐ Your test will run long enough to ensure that the quantity and type of data you collect can confidently answer the assumption(s) you're testing. If not, return to STEP 3 and reconsider your timeline.

- ☐ The prototype you have built is ready to do its job testing the particular assumptions targeted.
  If not, return to STEP 4 and refine your prototype.

- ☐ The prototype and research activities will gather the data you need.
  If not, return to STEP 4 and review the resources needed to ensure good data capture and collection.

- ☐ You have ensured that these tools will be used properly and consistently across testers.
  If not, return to the instructions in your research guide in STEP 4.

TEMPLATE 14

Concept Test History

Test number	Concept version	Keep	Eliminate	Change
date				
date				
date				
date				
date				
date				
date				

Log your progress in your **Progress Tracker**
[TEMPLATE 1, *page 82*]:

☑ Audit and pretest your design

....

☑ Run your test

....

☑ Share your findings

....

☑ Iterate your concept

....

Now you have some decisions to make about what comes next. Does your concept still look promising, and is your confidence in the accuracy of your assumptions high? It may be implementation time! But if you still have important remaining unknowns to address, you need to use what you've learned in this test to improve your concept, then repeat the cycle (starting with Step 2 and teeing up a new set of assumptions) to design your next test. However, if "make-or-break" assumptions were disconfirmed, it is likely time to table your concept (and go back to Step 1 and select a different concept to try).

Let the fun begin again!

Postscript

CONGRATULATIONS! You have come a long way since the beginning of this book. Now you know that experimentation is not only for scientists in lab coats with beakers. It's for you! As a changemaker, you're equipped with a powerful new set of tools to confidently explore the unknown. So, use your new experimental prowess to shine a light on brand-new ideas. Enjoy sharing this toolkit with others and seeing their confidence increase.

We are grateful to our partners at Nike, Whiteriver, PMI, and SWR for sharing their processes and learnings with us and allowing us to share them with you, our readers. Are you curious about what these intrepid experimenters are up to these days? We thought you might be . . .

Whiteriver Hospital's renovations have been completed and their new emergency room is hard at work.

At PMI, *Career Navigator* has become a successful new tool for their almost 700,000 members globally. One member commented: "It's about time! PMI has really listened to their members and created this awesome tool that is much needed for our profession!" Additionally, PMI's *Snippets* concept, launched under the name "PMI Picks," is also showing great engagement from members. As one shared: "The content is fantastic . . . it's clear, actionable, and I love that it's the perfect size to squeeze into my busy days, so I don't neglect my own development." We see how multiple tests and disciplined experimentation lead to a reward: offerings that those we design for love that also bring value to the organization.

For Nike, the story was different. Although *Easykicks* launched formally as Nike Adventure Club and generated multimillions in revenue, the company decided to discontinue the service after several years. Amid Nike's constant testing and launching of new offerings, Nike Adventure Club, although successful, did not make the cut.

At SWR, senior leadership remains excited about the continuing results they are seeing from their work with DK&A. The systematic step-by-step experimental approach is providing the organization with a clear pathway for innovation and change. Framing interventions with research from passengers is keeping everyone's feet on the ground. The safe space for co-creation with a diverse range of employees continues to drive optimism, momentum, and commitment to make things happen—and the board-level confidence built by test results is getting it all off the ground. Alan Penlington, SWR's customer experience director, believes that the best is yet to come: "We have an opportunity to showcase how to move the entire industry forward. There's a lot of change going on in UK rail—it's going to be a once-in-a-generation shift for the industry. We have the opportunity not just to float ideas, but to say instead, 'This is how we've done it. Let us get involved, let us help shape what the future is going to look like.'"

Regardless of where you find yourself in the experimentation process as this book ends, we hope that you think of this journey we've been on together as the development of a lifelong skill, not a one-off episode. In a world of uncertainty, treating everything you think you know as a hypothesis, challenging your own assumptions about what you believe to be true, should be a daily habit—and experimentation needs to be a way of life. We hope this book has prepared you to succeed at that bigger, more important journey and that our work together has given you the confidence to move your world in a new direction.

Happy experimenting!

Acknowledgments

WE WOULD LIKE to acknowledge the ongoing support of the UVA Darden Business School and the Batten Institute for Innovation & Entrepreneurship, as well as Innovate Carolina at the University of North Carolina at Chapel Hill. Without their funding and encouragement, this work would not have been possible.

Manjari Kumarappan and Emily Skywark, research assistants from the Department of Health Behavior at the UNC Gillings School of Global Public Health and graduate students in the Certificate in Innovation for the Public Good, were instrumental as researchers, creators, and reviewers.

Tim Ogilvie, founder of Peer Insight, as well as Clay Maxwell & Josh Clayton, Partners at Peer Insight, collaborated closely with Natalie Foley for a decade in developing and using much of this methodology.

A project of this nature demands co-creation with users and iteration of several prototypes. We would like to thank Karen Hold, Graham Henshaw, Art Brooks, and Danielle Lake for their great feedback on early prototypes that significantly shaped the direction of the materials.

Most important, we want to acknowledge Leigh Ayers, our amazing book designer, whose work reminds us every day of the magic of design done well.

Templates

TEMPLATES

TEMPLATE 1

Progress Tracker

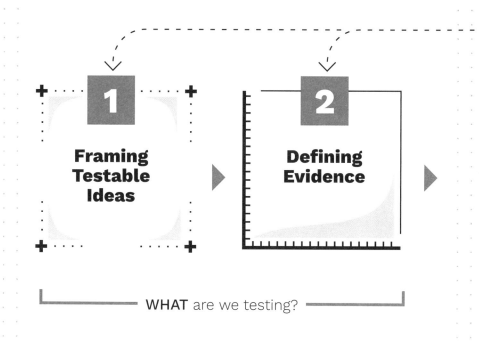

WHAT are we testing?

Step 1

☐ Frame a testable idea

☐ Prioritize concepts for testing

☐ Complete a Concept Snapshot

Step 2

☐ Surface critical assumptions

☐ Prioritize assumptions for testing

☐ Establish what constitutes evidence and identify targets

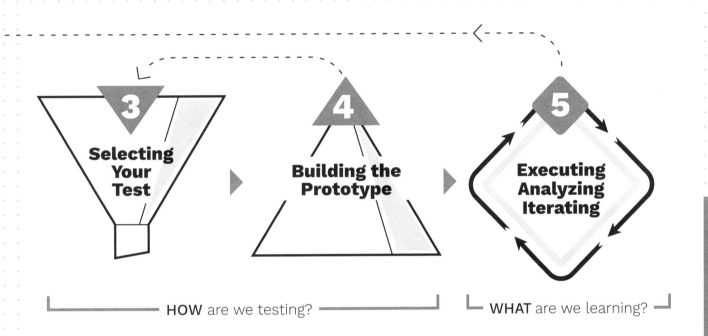

HOW are we testing? **WHAT** are we learning?

Step 3

- ☐ Choose between Say and Do tests
- ☐ Select a test type
- ☐ Design your test

Step 4

- ☐ Assess the level of fidelity needed
- ☐ Select prototype format
- ☐ Construct your prototype

Step 5

- ☐ Audit and pretest your design
- ☐ Run your test
- ☐ Share your findings
- ☐ Iterate your concept

TEMPLATE 2

Value/Effort Matrix

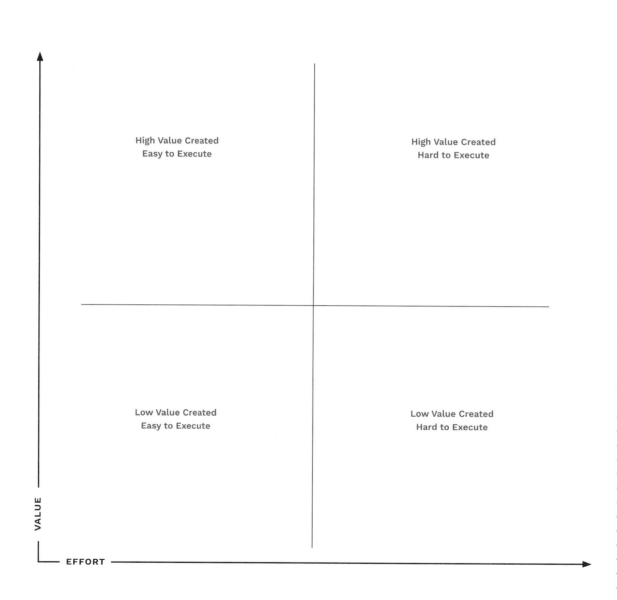

TEMPLATE 3

Concept Snapshot

Concept Name ....

	User 1	**User 2**
For (target user)		
Who want (unmet needs)		
We will offer (offering)		
That provides (benefits)		
Uniquely (differentiation)		

TEMPLATES

TEMPLATE 4

Storyboard

User experiences problem	User seeks solution	User discovers a better way

---

User tries out the new solution for the first time	User adopts the solution	User uses the solution again or refers others

TEMPLATE 5

Surfacing Assumptions

Concept Name ....

Desirability

- Users want it
- Users will choose it over other alternatives
- Users will pay for it

Feasibility

- We have the capabilities to produce the offering
- Users can access and use the offering
- We have the necessary partners

Viability

- We can scale both demand and delivery
- We can find these users affordably
- We can maintain competitive advantage
- The financials are sustainable

TEMPLATE 6

Prioritizing Assumptions

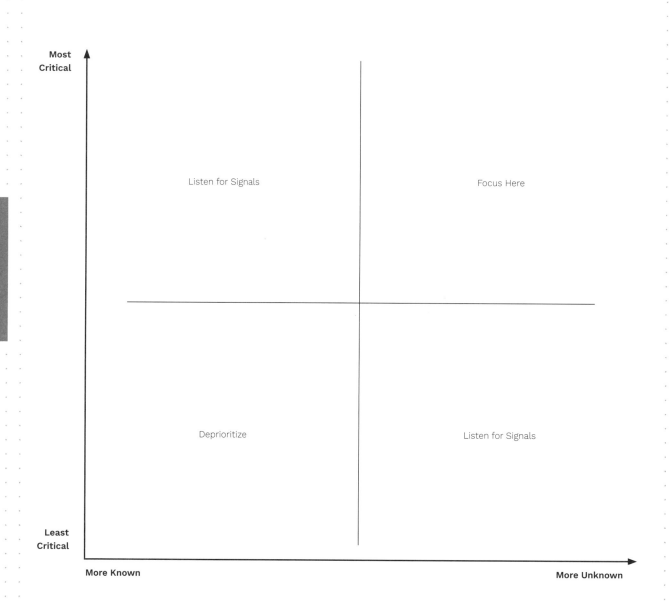

TEMPLATE 7

Assumptions to Evidence

Assumption	Evidence	Threshold Targets	Source

TEMPLATE 8

Data Sort

Assumption	EVIDENCE		
	What We Know	**Knowable with Existing Data**	**Knowable Only with Field Data**

TEMPLATE 9

Say/Do Continuum

Put an X or a dot along the continuum noting the thoughts of your team and stakeholders. The farther your dots are to the right, the more important Do data will be.

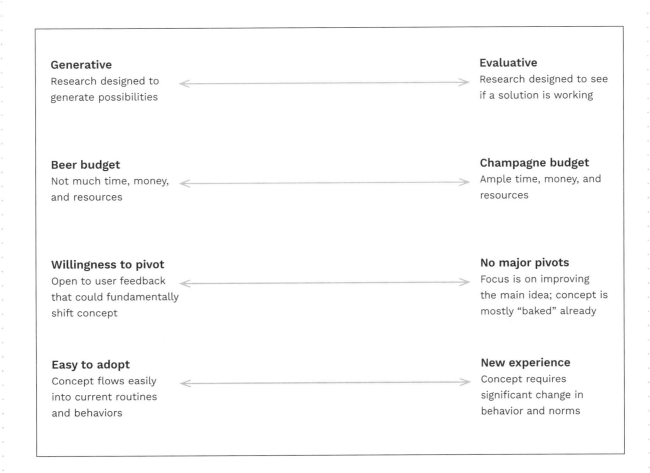

Generative
Research designed to
generate possibilities

Evaluative
Research designed to see
if a solution is working

Beer budget
Not much time, money,
and resources

Champagne budget
Ample time, money, and
resources

Willingness to pivot
Open to user feedback
that could fundamentally
shift concept

No major pivots
Focus is on improving
the main idea; concept is
mostly "baked" already

Easy to adopt
Concept flows easily
into current routines
and behaviors

New experience
Concept requires
significant change in
behavior and norms

TEMPLATE 10

Test Design Decision Flow

Q1 **Does data already exist to allow you to test a given critical assumption?**

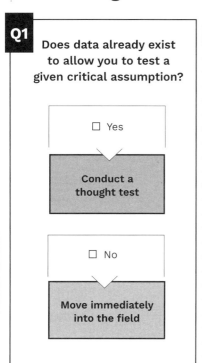

☐ Yes

Conduct a thought test

☐ No

Move immediately into the field

Q2 **Should you test a component or the whole concept?**

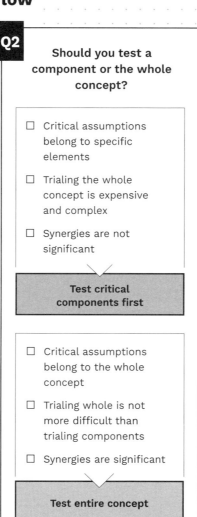

☐ Critical assumptions belong to specific elements

☐ Trialing the whole concept is expensive and complex

☐ Synergies are not significant

Test critical components first

☐ Critical assumptions belong to the whole concept

☐ Trialing whole is not more difficult than trialing components

☐ Synergies are significant

Test entire concept

Q3 **Do you need Say data or Do data?**

☐ Goal is to continue to learn about user needs

☐ Stakeholders have limited time, money, and resources

☐ Concept is evolving—pivots are fine

☐ Users can adopt new concept without changing routines

Select Say approaches

☐ Goal is to test the specific solution at hand

☐ Stakeholders will provide significant time, money, and resources

☐ Significant change in concept is unwelcome

☐ Significant behavioral change is required of users

Select Do approaches

TEMPLATES

Q4

What particular Say or Do test is most appropriate?

☐ Your goal is generative	☐ Your goals are evolving	☐ Your research goals are moving toward the evaluative	☐ Your research goal is evaluative	☐ Your goal is evaluative, with a specific solution to be tested
☐ Say data is acceptable	☐ You want to begin transitioning to Do data	☐ You want to gather arms-length Do data on interest	☐ Stakeholders expect Do data and will provide resources	☐ Do data is important
☐ You want flexibility to test components or whole	☐ You can locate your prospective users en masse at some locale	☐ You want to test specific value propositions	☐ You have confidence in your understanding of users' needs	☐ You want to test a full end-to-end concept
☐ Learning quickly and cheaply is important	☐ You have time and money to create stimulus and run stand	☐ You have money for ads and time to run ongoing tests	☐ You lack time and/or money to build a functional back end	☐ It is possible or even preferable to build a functional back end
☐ You don't have a lot of knowledge of users' needs and pain points				
Cognitive Walkthrough	**Lemonade Stand**	**Smoke Test**	**Simulation**	**Trial**

SAY APPROACHES ←——————————————————————————————→ DO APPROACHES

TEMPLATES

TEMPLATE 11

Test Digest

This document notes the key decisions you've made in designing a particular test before and after each of the five steps. Template 14 is where you'll track the key information for multiple tests.

1 + **2** **WHAT** are we testing?

Concept Name

Test Number

Assumptions
What are the make-or-break assumptions for this test?

1.

2.

REFER TO TEMPLATE 6

Evidence and Thresholds
Data to be collected during the test

Type of Evidence   Threshold Targets

REFER TO TEMPLATE 7

3 + **4** **HOW** are we testing?

Test Type
What type of test will you use to gather your data?

REFER TO TEMPLATE 10

Prototype
What prototype format will you use?
What does your prototype look like?

REFER TO TEMPLATE 12

Research Participants
Who are they and where will you find them?

REFER TO **RECRUITING PLAN**

Sample Size
How many will you recruit?

Time Frame
How long will you have to run the test?

Budget

Now that you have completed your test, time to compare your results with your targets, assess whether your assumptions are confirmed or disconfirmed, and reach conclusions about the future of your concept and whether you choose to iterate and test further, proceed as-is, or shelve.

5 **WHAT** are we learning?

Assumption Results
Data actually collected

Actual Results Compared to Target | Assumption Status

Assumption 1 was: ○ confirmed ○ disconfirmed

Assumption 1 was: ○ confirmed ○ disconfirmed

Assumption 2 was: ○ confirmed ○ disconfirmed

Assumption 2 was: ○ confirmed ○ disconfirmed

Insights
What else did you learn that is relevant?

Implications for Concept
How do these results and insights impact your concept? If iterating, what do you want to keep, eliminate, or change?

REFER TO TEMPLATE **14**

TEMPLATE 12

Prototype Format Selection Tool

To select your prototype format, use the template below. Put an X or dot along the continuum noting the thoughts of your team and stakeholders. The farther your dots are to the right, the higher the fidelity of your prototype needs to be.

How real does your prototype need to look? low fidelity ←————————————————→ high fidelity

Where is the concept in its development? early ←————————————————→ late

What assumptions are you testing? desirability ←————————————————→ feasibility and viability

How real does your prototype needs to feel? fake is fine ←————————————————→ authentic as possible

How does it need to function? not at all ←————————————————→ well

How much interactivity? tester can drive and support ←————————————————→ user drives unaided

What level of resources do you have to work with? little—needs to be cheap ←————————————————→ whatever it takes to make it feel and act real

How much time do you have? little—needs to be fast ←————————————————→ enough to follow whatever test plan calls for

TEMPLATE 13

Test Audit Checklist

☐ The Concept Snapshot for this idea specifies who you have designed it for, how it meets their needs, and why it is differentiated from other options already available.
If not, return to STEP 1 and refine your snapshot entries.

☐ The critical assumptions to be tested in this round are clear.
If not, return to STEP 2 and review the assumptions to be tested.

☐ The evidence you will gather aligns with the assumptions to be tested.
If not, return to STEP 2 and review the evidence needed and targets assigned.

☐ You have established clear threshold and aspirational targets.
If not, return to STEP 2 and establish these targets.

☐ You have selected the test type best suited to your situation. If not, return to STEP 3 and retrace your progress through the questions posed by the decision flow.

☐ You have recruited the right people to test with.
If not, return to STEP 3 and review your recruiting guide and instructions.

☐ Your test will run long enough to ensure that the quantity and type of data you collect can confidently answer the assumption(s) you're testing. If not, return to STEP 3 and reconsider your timeline.

☐ The prototype you have built is ready to do its job testing the particular assumptions targeted.
If not, return to STEP 4 and refine your prototype.

☐ The prototype and research activities will gather the data you need.
If not, return to STEP 4 and review the resources needed to ensure good data capture and collection.

☐ You have ensured that these tools will be used properly and consistently across testers.
If not, return to the instructions in your research guide in STEP 4.

TEMPLATE 14

Concept Test History

Test number	Concept version	Keep	Eliminate	Change
date				
date				
date				
date				
date				
date				
date				